Style Guide to Home Decor

& Furnishing

软装配饰风格 释义与解读

嶺南美術出版社

中国·广州

图书在版编目（CIP）数据

软装配饰风格释义与解读 = Style Guide to Home Decor & Furnishing ：汉英对照 / 深圳市艺力文化发展有限公司编. —广州：岭南美术出版社，2017.7

ISBN 978-7-5362-6144-0

Ⅰ.①软… Ⅱ.①深… Ⅲ.①住宅－室内装饰设计－汉、英 Ⅳ.①TU241

中国版本图书馆CIP数据核字(2017)第009342号

出 版 人：李健军
责任编辑：刘向上 张柳瑜
责任技编：罗文轩
特约编辑：李爱红 李 娟 胡 宁
美术编辑：陈 婷 蓝 梦

软装配饰风格释义与解读 = Style Guide to Home Decor & Furnishing
RUANZHUANG PEISHI FENGGE SHIYI YU JIEDU

出版、总发行：岭南美术出版社 （网址：www.lnysw.net）
（广州市文德北路170号3楼 邮编：510045）
经 销：全国新华书店
印 刷：深圳市汇亿丰印刷科技有限公司
版 次：2017年7月第1版
2017年7月第1次印刷
开 本：889mm×1194 mm 1/16
印 张：15
ISBN 978-7-5362-6144-0

定 价：320.00元

PREFACE 序言

Soft Decor Design, the Soul of Space

"Soft Decor Design", this concept is put forward in comparison to the traditional "hard" design, which means all the movable, flexible, and replaceable elements in a space, such as furniture, lighting, home textile, decoration and ornaments, etc. If we say hard interior design is the skeleton of a space, soft design will be the soul of it. As the design industry blossoms, soft interior design gradually bourgeons with the public increasingly recognising it. "Light finishing, and heavy decoration", a saying we always refer to, greatly assures the significance of soft decor design.

In the recent years, the style of soft decor design tends to be more simple and succinct – materials, structures and colours are all simplified and streamlined. Besides, more people have started paying attention to Asian culture, while focusing on the details of history and culture, also requesting highly for quality and comfort of space. Surely, there are people pursuing more "personalised" or tailored space, with the "mix-and-match" way, creating distinctive icons to make their home absolutely unique.

I have always had affirmative votes for soft decor culturally. Soft decor design is, instead of the simple accumulation of products, the cultivation of some sort of aura, and where a certain life style begins. By refining the functions, such design enables people to live more comfortably and conveniently. Not just about purchasing products, an interior soft decor designer is required to have unique understandings on space design, rich product resources, and ability of selecting and matching all different soft interior elements, so that they can further optimise the design with the help of the concise cognition of the space, which is also a direct representation of the delight in life as well as the literacy in culture.

Soft decor design is a specialised skill and profession. If you would like to have more professional and in-depth understandings on this industry, I would recommend you to start from this book.

Jiang Xiaolin
Co-Direction Interior Design

软装，空间的灵魂

软装设计的概念是相对于传统硬装设计而言的，一个空间里，所有可移动替换的元素都是软装，如家具、灯饰、布艺、饰品摆件等。如果说硬装是一个空间的骨骼，那么软装就是这个空间的灵魂。随着设计行业的蓬勃发展，软装配饰也逐渐走进大众的视野，得到了越来越高的重视，我们常常说的"轻装修，重装饰"就是对软装设计地位的极大肯定。

近两年，软装配饰的风格逐渐趋于简洁，无论是材质、结构还是配色，都在不断地做减法。越来越多的人也开始关注东方文化 ，在注重历史文化细节的同时，也对空间的质感与舒适度产生了更高的要求。当然，也有一部分人追求更加"个性化"的定制空间，通过混搭的手法去打造独一无二的印记。

我一直倡导文化软装，软装不是一些产品的堆积，是一种气质的营造， 一种生活方式的开始，软装细化了功能，让生活更舒适方便。软装配饰不是简单的产品购买，一名软装设计师需要对空间设计有自己的理解，掌握丰富的产品资源，具备选择与搭配软装元素的能力，通过对空间的精准理解而进行优化，是生活情趣和文化素养的直接体现。

软装设计是一门专业，如果你想要对这个行业有更专业、深入的了解，那么不妨从这本书开始。

姜晓林
共向设计创始人

CONTENTS

目录

ADVANCED MINIMALISM

高级简约

Simplicity Love
简爱

"From tomorrow on, I will be a happy man;

Grooming, chopping and traveling all over the world;

From tomorrow on, I will care about foodstuff and vegetable,

living in a house towards the sea with spring blossoms."

Does Haizi's poem intrigue you? Using the burning preference and passion to avoid the hustle and bustle of city life, and choosing a romantic coast to live. Seeing the ebb and flow every day and experiencing the ups and downs of mood. Let the sunshine fill the room, and decorate it with odd-shaped shells and freshly picked buds, delicate glass bottles and jars, a coir carpet and coverings hand-made with white cotton and linen fabric. Carefully preparing a delicious meal, just like the poem — "skillfully mincing meat as smooth as snow, with brilliant seasoning making the dishes as crowd-pleaser", cooking with gratitude towards the person and the ingredients, and by doing so, you could create the slow life of your own. Changes in life can be easily made when taking on a different perspective to it, the whole world begins to change once you look at it differently, like the mindset described in Haizi's poem — I face the sea, with flowers blossoming in the spring breeze.

Some say happiness is money that can't be used up, and being able to laugh everyday; some say happiness is when the boss allows you to have a few paid days off. Obviously, the former rises from the lack of money, while the later comes from the lack of time. But when you have enough life experiences, money and time, you would realize that, happiness is simple – slow down your pace and live on the basis of life, simplicity is the best, and that's how life should be, simple and warm with some occasional surprises.

Wonder if you have such experience – In a slightly humid but comfortable afternoon, you take a glimpse into the wall corner and find a small handful of weed just grow out of it, they are still fresh and soft, "should I get rid of them", you think, but you then hesitate: "that is a life, a new-born life, I should leave it be". And with changing seasons, they are getting greener and stronger. When the winter comes, you get worried and see if they get frozen, upon your touch, although feeling a bit dry, they are soft and still standing against the cold and waiting for the coming year, when they could thrive.

To some extent, our life is similar to those weeds — we can't choose where we come from as well as the unpredictable toughness along the way, the only thing we could determine is the way we live our life, we better keep our head down, get our hands dirty and focus and stop complaining, learn the necessary skills, and that's it, keep your life simple, instead of worrying and shouldering too much pressure and desires. It may be hard to make our life easier, but we can always make it simpler by keeping a clear vision on your true goal and doing the necessary things to achieve it. So, keep going, keep your life simple.

“从明天起，做一个幸福的人；

喂马、劈柴，周游世界；

从明天起，关心粮食和蔬菜，我有一所房子，面朝大海，春暖花开……”

海子的诗有没有让你奋而跃跃欲试，要用燃烧的偏好与激情，避开车马喧嚣，择一湾浪漫的海边住下，每天坐看潮起潮落，收录心境的起伏得失；收捡奇形怪状的贝壳，采摘鲜艳欲滴的花蕾；为洒满阳光的一室空间，拾掇精致的玻璃瓶罐、低调内秀的质朴饰品，铺上弥漫一地的椰壳纤维地毯、手工的白色棉麻织物；专心致志地下厨做上一顿珍馐美食，“无声细下飞碎雪，放箸未觉全盘空”把人与食物的缘分揉融在心……就这样创造并享受着属于你的慢生活。物随心转，境由心造，这种生活如诗般美好，正如海子理想的面朝大海、春暖花开般幸福。

有人说，幸福是有钱任意花，天天笑哈哈；有人说，幸福是跪求老板放自己几天“解压”假期。显而易见的是，前者大多属于缺钱，后者大多缺时间。唯有握住岁月累积而得的经历与体悟，达到既不缺钱也不缺时间的完满状态，此刻你眼中的幸福却会是很简单——让生活慢下来，回归生活本真，过上简单的生活。这就是这一章节的主题——高于生活之上的简约主义生活方式，试图让生活回归原点，舍去繁复的枝节，找回生活应有的温度。

你有没有过这样的经历，在一个春日微湿的聊赖午后，不经意地瞥见四方墙壁的角落里，竟不知何时长出一株嫩弱的小野草，犹豫着要不要把这水泥缝中的小东西清理掉，心里最柔软的部分被触动了一下，这也是一个小生命啊，还是留着吧。挨过春天、夏天、秋天的日子，小野草渐渐地长高了，变绿了，还冒出几簇新芽，为这一隅的角落，增添了一抹倔强的生机。接着冬天到了，你怕小野草在这阳光很少的角落里冻死，特意用手摸了摸它的茎叶，韧韧的，没有干枯的迹象，这才放下心来，相信它一定会在冬天的寒冷中挺住，再在新的年头继续生长。

人生也一如这小野草吧。出身无法选择，际遇不容躲闪，唯一能够左右的就是学会生活，学会成长，没有怨言地生发，努力地拔高向上。不要再为自己的心负载更多的冗余，努力找回忘记了的理想和初心，因为“心简单，这个世界就不复杂”，懂得单纯的生活更可贵，所有延伸于生活之外的欲望都是生命的负累。简单，才是生活的本质；复杂，是心蒙尘后的幻象。卸掉心的冗余，清扫心灵的垃圾，笃定生活的信念，始终保持昂扬的姿势，时刻让心充满阳光，一路前行，一如既往。

Simplified but not simple, life back to its nature.

——简约而不简单，生活回归本质

IMAGE © GLOBAL VIEWS

图片来自 GLOBAL VIEWS 品牌

IMAGE © GLOBAL VIEWS
图片来自 GLOBAL VIEWS 品牌

Advanced Minimalism Proposition
高级简约主张

Minimalism comes from Western Modernism in the early 20th century, its founder is Walter Gropius. To promote function as the first principle, Gropius made furniture modeling fit for the production line. To promote minimalism in the architectural decoration, it simplified them including the design elements colors, lighting, raw materials to a minimum extent which improved the requirements of colors and material texture. It tried to achieve decorative effect of doing more with less. Minimalism developed on fundamental of rebellion against the trend of retro and Minimalist aesthetics in the mid-80s, and began to integrate into the field of interior design in the early 90s. This simple form of expression meets the needs of people. For example, emotional, instinctive and rational needs of space environment. It is gradually being the concern of the international community today.

Minimalism is not simple. It derived from careful consideration and innovation, rather simply "pile up" or "placed." Concise but not simple, it could be the best portrayal of modern minimalist style. No more decoration, it pays attention to modest proportion of modeling, emphasizing bright appearance, simple lines and functional design, and color in strong contrast or elegant pleasure. It reflects the modern fast-paced, simple and practical, but vibrant life. Because of the simple line and less decorative elements, modern style furniture needs the perfect soft decoration, in order to show beauty. Accessories in modern minimalist style are the most eclectic. Some ornaments with simple lines, unique design and even very creative and personalized style may be treated as modern minimalist.

简约主义来源于20世纪初期的西方现代主义，创始人是瓦尔特·格罗皮乌斯（Walter Gropius）。提倡功能第一为原则的格罗皮乌斯，提出了适合流水线生产的家具造型，在建筑装饰上提倡简约，将设计的元素、色彩、照明、原材料简化到最少的程度而提高了对色彩、材料的质感的要求，致力达到以少胜多、以简胜繁的装饰效果。简约主义从80年代中期对复古风潮的叛逆和极简美学的基础上发展起来，90年代初期，开始融入室内设计领域当中。这种以简洁的表现形式来满足人们对空间环境那种感性的、本能的和理性的需求，正逐渐受到当今国际社会关注。

简约不等于简单，它是在深思熟虑后经过创新得出的设计和思路的延展，不是简单的"堆砌"和平淡的"摆放"。简洁但不简单，或许是对现代简约风最好的写照。没有过分的装饰，讲究造型比例的适度、强调外观的明快、线条简约、功能性设计，以及色彩或对比强烈或淡雅宜人，体现出现代生活快节奏、简约和实用，但又富有朝气的生活气息。由于线条简单、装饰元素少，现代风格家具需要完美的软装配合，才能显示出美感。现代简约风格饰品是所有家装风格中最不拘一格的一个。一些线条简单、设计独特甚至是极富创意和个性的饰品，都可以成为现代简约风格家居中的一员。

Characteristic

Minimalism emphasizes practicality. It also stresses the individuation and abstraction of indoor space form and component, and pursues the depth and precision of material, technology and space, and reflects the concise and bright sense of times and abstract beauty.

Lines

Minimalism often uses geometric structure, composited of curve and asymmetric lines. Some lines are soft and elegant, and some vigorous and full of sense of rhythm. The whole three-dimensional form integrates into methodical, rhythmic curves. The use of simple structure and beautiful shape brings pleasure and leisure, which is the pursuit of a modern psychological comfort.

Pattern

Walls, railings, window frames and furniture are with such pedicels, buds, vines, insect wings and a variety of natural beauty, wavy shape patterns. Compared with the traditional style, modern minimalism with the most straightforward decorative language clarifies the home space to create the atmosphere, and then gives the space personality and calmness, but also presents the avant-garde, unfettered feeling.

Color

Minimalism avoids complex color scheme but different from the absolute black and white, it stands out in an implicit way. To achieve that, it requires a simplified colour selection from colours of similar shapes, or the ones in contrast.

Material

The quality and texture of Minimalist deco is the new luxury. Instead of those strong and aggressive use of lines and materials, the modern Minimalism takes on materials like steel grass, stainless steel. Also, traditional craftsmanship is involved, the use of crafted glass, ceramic and iron work in their pure and elegant forms, adding a sense of art to your space.

特征

注重实用，强调室内空间形态和构件的单一性、抽象性，追求材料、技术、空间的表现深度与精确；体现简洁明快的时代感和抽象的美感。

线条

多采用几何结构，由曲线和非对称线条构成，线条有的柔美雅致，有的遒劲而富于节奏感，整个立体形式都与有条不紊的、有节奏的曲线融为一体。利用简单的结构及优美的造型给用户带来愉快安逸，这是现代人所追求的一种心理安慰。

图案

墙面、栏杆、窗棂和家具上带有如花梗、花蕾、葡萄藤、昆虫翅膀以及自然界各种优美、波状的形体图案等。与传统风格相比，现代简约用最直白的装饰语言体现家居空间营造的氛围，进而赋予空间个性和宁静，又能给人带来前卫、不受拘束的感觉。

色彩

既没有复杂的色彩搭配，又有别于绝对黑白的古板，却可以在不露痕迹的风情间含蓄地绽放着独特光芒。要达到简约风格，同时施展个性，就需要色彩跳跃和律动。大量运用苹果绿、深蓝、大红、纯黄等高纯度且对比强烈的色彩，将空间的节奏感融入到室内装饰中。

材质

不需要夸张的噱头，足以用纯粹材质博得当之无愧的奢华主角。金属材质的装饰品是体现简约风格最有力的手段。现代简约派金属灯大多使用钢化玻璃、不锈钢等新型材料作为辅材，也是现代风格家具的常见装饰手法。大量使用铁制构件，玻璃、瓷砖以及铁艺制品、陶艺制品等均以简洁的造型、纯洁的质地、精细的工艺为其特征，竭力给室内装饰艺术引入新意。

1

GLOBAL VIEWS

环球视野

IMAGE © GLOBAL VIEWS
图片来自 GLOBAL VIEWS 品牌

IMAGE © GLOBAL VIEWS
图片来自 GLOBAL VIEWS 品牌

This clam but not boring space takes large white area as backdrop, which is enriched with furniture and accessories of gold colour and various shapes of beige.

客厅运用了大面积白色搭配金黄、米白色系的家具和配饰，不仅丰富了整个空间气氛，更营造了恬淡素雅的空间氛围。

The corner of living room chooses warm colored wood as main material. Wall painting and desk decorations have become the focus.

客厅一角选择温润的木质材料为主，而墙上的壁画和书桌上的装饰成了焦点。

1.1 LIVING ROOM
客厅赏析

IMAGE © GLOBAL VIEWS
图片来自 GLOBAL VIEWS 品牌

IMAGE © GLOBAL VIEWS
图片来自 GLOBAL VIEWS 品牌

IMAGE © GLOBAL VIEWS
图片来自 GLOBAL VIEWS 品牌

Bright and spacious sunroom makes people more physically and mentally pleasant. Black cantilever chairs and floor echo, but also cleverly become the "highlight" in light-colored space:

With its Medieval origin, the chair's new suspension system supports the leather-clad seating area with polished stainless frame, creating a hammock-like seating experience. This comfort is also enhanced by the selection of short-haired leather, adding a sense of classic.

宽敞明亮的日光室让人的身心倍感愉悦，黑色悬臂椅与地板交相呼应，也巧妙地成为这个浅色系空间的“亮点”：

中世纪新式悬吊设计，新颖美观且极具舒适感；宛如吊床一般的皮革座椅由打磨过的不锈钢架作为支撑，坚固而平稳；短毛皮革带来细腻的触感，不论是纹理还是颜色都体现出设计之经典。

WARI
Lords of the Ancient Andes

IMAGE © GLOBAL VIEWS
图片来自 GLOBAL VIEWS 品牌

Decorated by black branches, the cabinet feels calm with a sense of design. And with red coral pattern carpet and bright-colored accessories, the overall colors are lively and full of passion.

黑色树枝装饰柜沉稳中又不乏设计感，结合红色珊瑚花纹地毯和亮色配饰，整体颜色华丽生动而富有热情。

For color selection, white, ivory and grey tone colors are widely used across the space with blue and green as embellishment, highlighting a sense of tradition and elegance.

在色彩运用上，以白色和象牙色为主色调，蓝色和绿色点缀其中，并大量使用灰色系，凸显传统与大气。

The living room is mainly in golden yellow, including the tea table and side table, making the whole room bright. The lamping bracket is fully covered by fan-shaped brass leaves, and its branch-shaped appearance adds a little fun to its classical style. The cocktail table in the living room center is also unusual, with the smooth dumb white marble table and four long spike-like table legs, while the two extended nails that link across four corners making the table more stable. The small tea table next to the sofa is extraordinary, which is surrounded by the steel branches in casual. Braches are also intertwined by themselves, supporting the dumb white marble table. The entire appearance makes people thinking of a pair of tango dancers.

客厅以金黄色为主，不论是落地灯还是茶几、边桌，统统带有金色元素，使整个室内更加光亮。扇贝形状的黄铜叶子犹如鳞片一般覆盖在落地灯的支架上，树枝一般的造型让古典多出一分趣味；客厅中央的鸡尾酒桌也是不走寻常路，光滑的哑白色大理石桌面搭配四根长钉状桌腿，还有两根加强版长钉交叉连接四角，使桌子更加稳固；沙发旁的小茶几更是独具特色，随意环绕的铁质枝干犹如一个旋涡，又如同相互缠绕的树枝，支撑着顶部的哑白色大理石桌面，整体造型让人不禁联想到一对探戈舞者正在翩翩起舞。

The colors of furniture such as carpet, sofa and pillow are harmonious and vivid. There is no lack of stability in the changes, and the scenery outside the windows has become indispensable embellishment in the space.

地毯、沙发与靠枕的色彩搭配和谐中有律动，稳重中又不乏变化，而落地窗外的景致也成了空间中必不可少的点缀。

IMAGE © GLOBAL VIEWS
图片来自 GLOBAL VIEWS 品牌

1.2 DINING ROOM
餐厅赏析

With brown as the main color of restaurant, the clever use of similar color scale, and elegant green planting flowers, make the living room both decent and elegant.

以棕色为餐厅主色调，相近色阶的巧妙运用，搭配雅致的绿植花卉，使得整个客厅既得体又优雅。

The leaves will reflect the warmth of the golden glory, when the light is lit. The jagged dining table is inspired by Global Views' best-selling products-Serrated Wall Table. The zebra-like woodwork looks like a chessboard. The jagged base acts like an accordion. More specifically, the table itself is two layers, can be opened on both sides, after the extension can accommodate 12 people dining at the same time.

餐厅的锯齿形餐桌灵感来源于 Global Views 的热销产品锯齿形靠壁桌，横竖相交的斑马木纹看起来就像是西洋棋盘，锯齿形的底座宛如一个手风琴，两者结合形成了这款别致的餐桌。更为特别的是，餐桌本身为两层，可以将两边拉开，延伸之后能够容纳 12 人同时就餐。

餐厅的主色调则为棕色，餐桌的花形底座和透明玻璃桌面避免了沉闷，象牙色与棕色相间的餐椅显得高贵优雅，由坚固的玻璃管加热弯曲而成的 U 形玻璃烛台与玻璃桌面呼应，多个烛台组合成为餐桌上一道美丽的风景。靠墙而立的日内瓦玻璃橱窗以美国白栎木作为框架，三面玻璃外壁和镜子背景以及四块可调节玻璃板方便展示橱窗内的摆设，两盏顶灯体现出设计师的细致周到。

The dining room above uses brown as main color. To add a bit of lightness to the room colored with brown, lively shape and glass material are added to its furniture — the dining table with transparent glass top and flower-shaped base matched with a few U-shaped glass candlesticks; The wall-sided Geneva glass cabinet, framed by American white oak, has 3 transparent glass sides, and the four glass panels inside can be rearranged to suit your display need, with the mirror back and the built-in lighting, the cabinet is ideal for showcasing your favourite collection.

IMAGE © GLOBAL VIEWS
图片来自 GLOBAL VIEWS 品牌

1.3 STUDY ROOM
书房赏析

The study is elegant and ivory white-based, with carpets, cushions and decorative flowers to increase the temperature. Hollow decorative desk is mainly made of broadleaf hardwood, and there are very exotic zebra decorative panel, auburn surface coated with varnish and hand-waxed as a protective layer. The desk is equipped with a large drawer for stationery, such as pencils, adding functionality to its design.

上面的书房以文雅的象牙白色为主，地毯、靠垫及装饰花朵增加温度。镂空装饰写字桌主要由坚固的阔叶硬木制作而成，还有极具异国情调的斑马纹装饰面板，赤褐色表面涂以清漆并手工上蜡作为保护层。写字桌配有一个专门用来放置铅笔等文具的大尺寸抽屉，彰显出设计的细致周到。

A muted and neutral color scheme is used for the study below — the white and ivory used in the the furniture and decoration match well with that of the tricolour flower-patterned wallpaper. The heavily brown-colored desk is lightened by the transparency of crystal bulb-shaped base of the desk lamp, and together they become the special feature of the study. The curved lines and texture of the silk lampshade also help to soften the image of the space.

书房整体采用淡雅色系，大面积使用的白色和象牙色与灰绿白三色相间的花朵壁纸相映成趣，古典球茎水晶灯使稳重的棕色书桌不再沉闷，透明的灯身与底座成为书房最特别的存在，8 英寸的圆柱形灯罩以象牙色丝线覆盖，为书房增添了一抹柔和，也使其成为当之无愧的焦点。

The bedroom is painted by a warm color, yellow. Both bedding and lamp are designed to fit with the owner's comfortable sleep. The classical and innovative lamp can be described as a highlight of the bedroom. This lamp has a solid brass and vase shape body, with the quadrate pedestal made by enamel iron. Its body was twined by flowers, creeping to the top. The lampshade is also in bronze, aiming to be consistent with the body, due to the wonderful combination between the elegant bronze and classic black. Besides, the collocation by the rounded lampshade and angular square is decent.

卧室加入温馨黄色，无论是床品还是台灯都考虑到主人的舒适睡眠。这款极具古典特色又富设计新意的台灯可谓卧室的一大亮点：坚固的黄铜质地花瓶造型灯身，底座采用珐琅钢铁铸造成方形；灯身以花枝缠绕，蔓延至台灯顶部，灯罩同样配以古铜色，与灯身呼应——典雅的古铜色与经典黑色相得益彰，圆润的球形与硬朗的方形搭配得当。

1.4 BEDROOM
卧室赏析

IMAGE © GLOBAL VIEWS
图片来自 GLOBAL VIEWS 品牌

To help relax, a wide range of warm orange tone colors are applied to the bedroom. The bathroom is connected to the bedroom, and designer is bold to keep the two walls of the floor to ceiling windows. Concise and neat style bathtub was placed by the window. The owner can enjoy the views outside while bathing. The sculpture like two arms staying together is a stool made of Portuguese reinforced ceramic, the two open hands may look fun but also provide comfortable seating. The craftsmanship involved makes it an artwork standing alone, or you could follow the photo on the right, using it to hold your toiletries, and that works brilliantly.

卧室作为休息睡眠区域，暖橘色调被大范围运用。与卧室相连的浴室，设计师大胆保留两面墙的落地窗，造型简洁利落的浴缸被安置在窗边，主人可以边沐浴边欣赏窗外的美景。同时出现在浴室和阳台的哑白色双手托起陶瓷凳，是由葡萄牙工匠运用加固陶瓷烧制而成，坐在上面平稳且舒适，优雅的雕刻工艺使它单独放置时也仿若一件完美的艺术品。你还可以效仿设计师的创意，将它放置在浴缸旁，用来收纳毛巾和洗浴用品，也不失为一道独特的风景。

IMAGE © GLOBAL VIEWS
图片来自 GLOBAL VIEWS 品牌

1.5 PRODUCT DISPLAY
单品赏析

IMAGE © GLOBAL VIEWS
图片来自 GLOBAL VIEWS 品牌

DwellStudio Designers Collection

DwellStudio 设计师系列产品

02 Golden Palm Leaf Mirror

The palm-leaf-formed mirror frame is crafted with iron, on which there is shinny golden paint.

金色棕榈叶镜子

铁制框架形成独特的棕榈叶造型，金色涂漆使整面镜子仿佛闪闪发光。

01 Malachite Pattern Decorative Plate

Malachite pattern decorative plate is made of flown glass with malachite pattern. Standing alone, it's stunning enough, whether used for decoration or as a tableware.

孔雀石花纹绿色装饰盘

熔融玻璃打造，配以孔雀石花纹。如此美丽的单品，不论是用于装饰还是当作餐具，都足以令人惊艳。

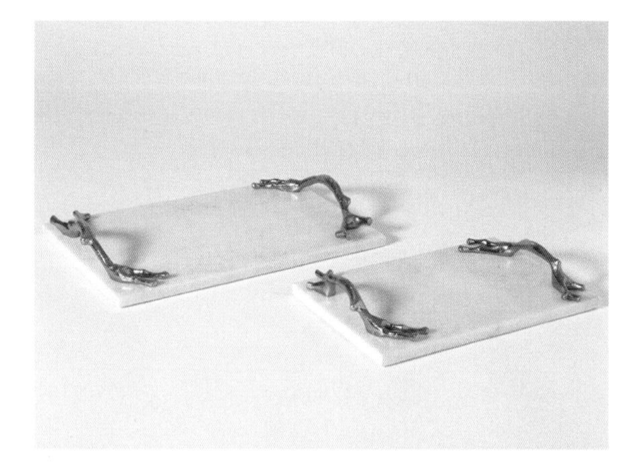

03 Brass Branches White Marble Tray — Large / Small

Solid brass is carved into a branch shape, mounted on a white natural marble plate. The dazzling handles, together with the marble plate, form a lovely chic tray.

黄铜树枝白色大理石托盘（大 / 小）

坚固的黄铜被雕刻成树枝的形态，镶嵌在白色天然大理石板上，金色的手柄显得更为亮眼，最终生成这款无比完美的别致托盘。

04 Silhouette Figure Table Lamp

Colored with fresh and elegant tones, the lamp takes Portugal-made ceramic portrait figure as lamp body, transparent glass base and grey nickel-plated lampshade.

人物侧影台灯

葡萄牙特制陶瓷人像作为灯身，灰色镀镍灯罩，透明玻璃底座，色系搭配清新淡雅，简约而不简单。

05 Light Blue Chamomile Ceramic Jar — Large / Small

The shape is classical. It is also functional. Beautiful light blue active glaze is soft and fresh.

浅蓝色甘菊陶瓷罐（大 / 小）

造型古典，兼具实用功能，漂亮的浅蓝色活性釉面柔和而清新。

06 Olive Green Carambola Bowl

The carambola bowls have elegant lines and fine details, which are used to pay tribute to a long-history and exquisite-craft Portuguese ceramic. When placed alone, it is a unique decoration; After putting fruits in, it turns into a beautiful artistic fruit plate.

橄榄绿杨桃碗

这组杨桃碗线条优雅，细节精致，用以向历史悠久且工艺精美的葡萄牙陶瓷致敬。单独摆放，它是一款造型独特的别致装饰；加入各式水果，它可变身为极具艺术色彩的创意果盘。

08 Matt White Carambola Box

The carambola series have elegant lines and fine details, which are used to pay tribute to a long-history and exquisite-craft Portuguese ceramic.

哑白色杨桃盒子

优雅的线条，精美的细节，这套杨桃系列是为向年代悠久且质地上乘的葡萄牙陶瓷致敬。

07 Babylon Floor Lamp

Taking the same central axis, a number of circular parts stacking up and down to form the lampshade of this floor lamp, the symmetrical design resembles the legendary tower standing wildly in field.

巴比伦落地灯

以一根中轴为核心，多个同心圆环绕而成，大胆的圆柱形对称设计使台灯看起来多了一份高耸的狂野。

IMAGE © GLOBAL VIEWS
图片来自 GLOBAL VIEWS 品牌

09 The White Marble Spike Table

With a golden appearance and white marble desktop, it is elegant. Three iron spike-type table foots are in simple line, making the table solid and stable.

白色大理石长钉桌

金色的外表配以白色大理石桌面，高贵典雅。三根铁质长钉形桌脚线条简洁，稳固又坚实。

10 Ivory Wayne Chair

Its erected high back, with handrails at the convergence of the smooth arc, and thick cushion, mean that every detail regards to the comfort of experience. With walnut chain legs, medieval wind is blowing.

象牙色韦恩椅子

直立的高耸靠背，与扶手衔接处的流畅弧度，以及厚实的软垫，每一处细节无不考虑到舒适体验。搭配胡桃木椅脚，中世纪风扑面而来。

11 Black / Ivory Wood Pallet

It uses medium density fiberboard as the main body. Its surface is cocoa wood decorated with varnish. Two small trays can be embedded in the large tray. Concise lines and the classic shape are never out of date.

黑色 / 象牙色木托盘

中密度纤维板做主体，表层是可可木饰以清漆，两个小号托盘可嵌入大号托盘当中。线条简洁，经典的造型永不过时。

12 White Leather Box With Handle

Full leather package and handle with nickel-plated stainless steel are noble and refined. Box body with linen lining, is your favorite collection for valuables.

白色带柄皮革盒子

全皮革包裹，手柄搭配镀镍不锈钢，显得高贵而精致。盒身以毡布作为内衬，是您收藏贵重物品的不二之选。

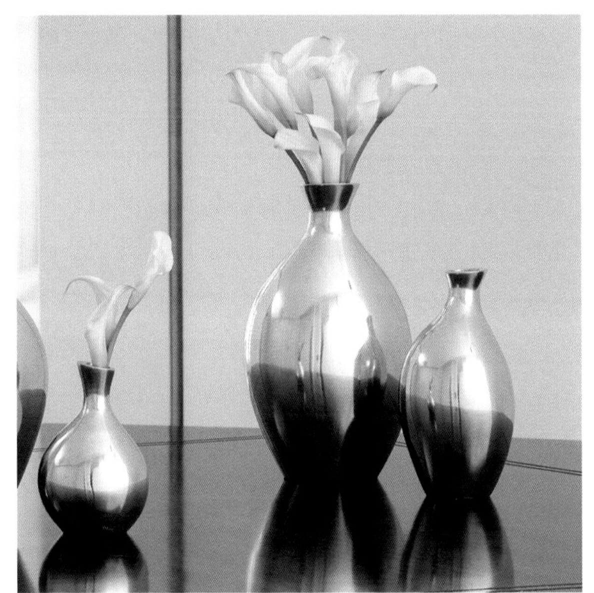

13 Aluminum White Vase — Large / Medium / Small

Aluminum vase presents silver plated appearance. It shows fresh taste when plugs a few branches of calla.

铝白色花瓶（大 / 中 / 小）

铝制花瓶换成镀银外观，插上几枝海芋，多了一份清新的味道。

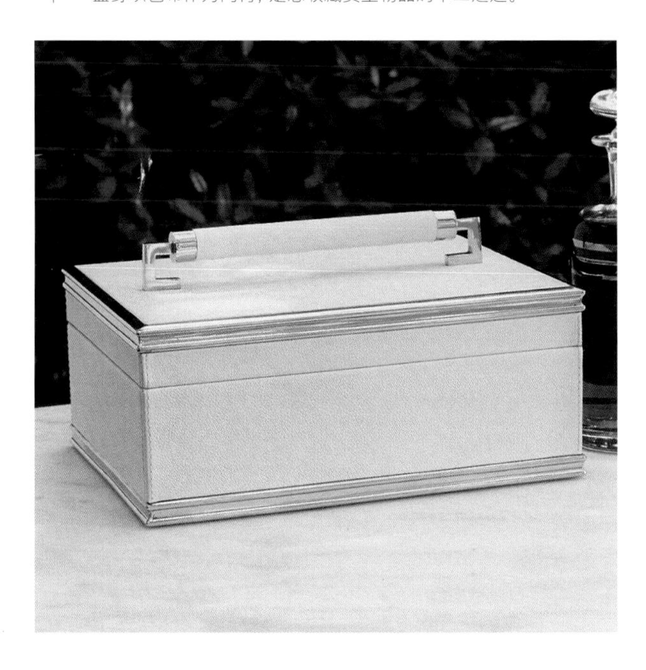

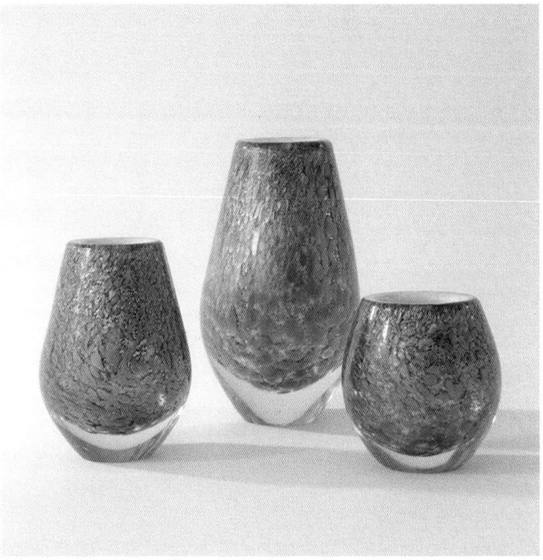

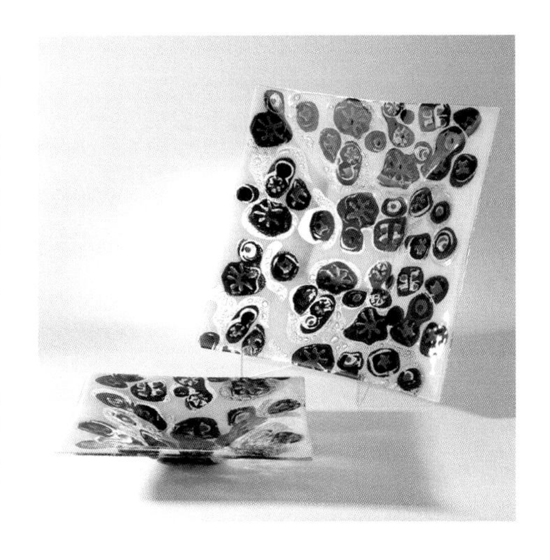

15 Copper Mesh Glass Vase — Large / Medium / Small

Glass vase with double-layer structure, is very strong in fact. In addition to metallic color dot-like texture at inner, it is inlaid with real metal debris. The combination of them forms incredible texture.

红铜色网状玻璃花瓶（大 / 中 / 小）

玻璃花瓶采用双层结构，实则十分坚固，内层除了金属色斑点状纹理之外，还镶嵌有真正的金属碎片，两者结合，形成不可思议的纹理。

14 Blue Amoeba Glass Bowl — Large / Small

Early porcelain became the source of inspiration for the glass bowl. The landslide effect of the flown glass made it a unique shape, and especially the amoeba pattern made it even more unique. The product has passed through the food safety test, which is not only a decoration, but also chic tableware.

蓝色变形虫玻璃碗（大 / 小）

早期瓷器成为这款玻璃碗的灵感来源，熔融玻璃制造的滑坡效果形成独一无二的形状，变形虫图案使其更显独特。这款产品已通过食品安全测试，因而不仅可以作为装饰，也可当成一件别致的餐具。

IMAGE © GLOBAL VIEWS
图片来自 GLOBAL VIEWS 品牌

2

OMENIA

欧米亚

IMAGE © OMENIA
图片来自 OMENIA 品牌

Fashion

2.1 LIVING ROOM

客厅赏析

IMAGE © OMENIA
图片来自 OMENIA 品牌

The colors of metal furniture grey walls and brown wood flooring are harmonious, and also enhance the sense of dynamic space.

金属家具与灰色墙面及棕色木地板颜色搭配协调，又提升了空间的跃动感。

IMAGE © OMENIA
图片来自 OMENIA 品牌

IMAGE © OMENIA
图片来自 OMENIA 品牌

Orange, brown and wood color, these warm colors help heat up the space from the visual. Abstract pattern of painting as a decoration makes up the wall of emptiness. Schema and tone are based on the main home style, to create a consistent indoor atmosphere.

橙黄、褐色、原木这些暖调的色彩从视觉上帮助空间升温，抽象图案的画作作为点缀弥补墙面的空荡，图式与色调的选择均依主体家居风格而定，营造室内氛围的吻合一致。

IMAGE © OMENIA
图片来自 OMENIA 品牌

There is light, bright wide space, and full of color, with rich patterns. There is temperature and reminiscent. Matching with cotton wool carpets and other objects, metal furniture has rich texture and its cold smell of the metal itself is balanced.

轻盈、明亮的宽阔空间，饱满的色彩、丰富的纹样，有温度、有联想。以棉毛地毯等物件搭配金属家具，既丰富了质感，又能平衡金属本身的冰冷气息。

IMAGE © OMENIA
图片来自 OMENIA 品牌

2.2 HALLWAY DISPLAY
玄关赏析

The color of metal furniture with black walls and brown wood flooring is harmonious, and also enhances the sense of dynamic space.

金属家具与黑色墙面及棕色木地板颜色搭配协调，又提升了空间的跃动感。

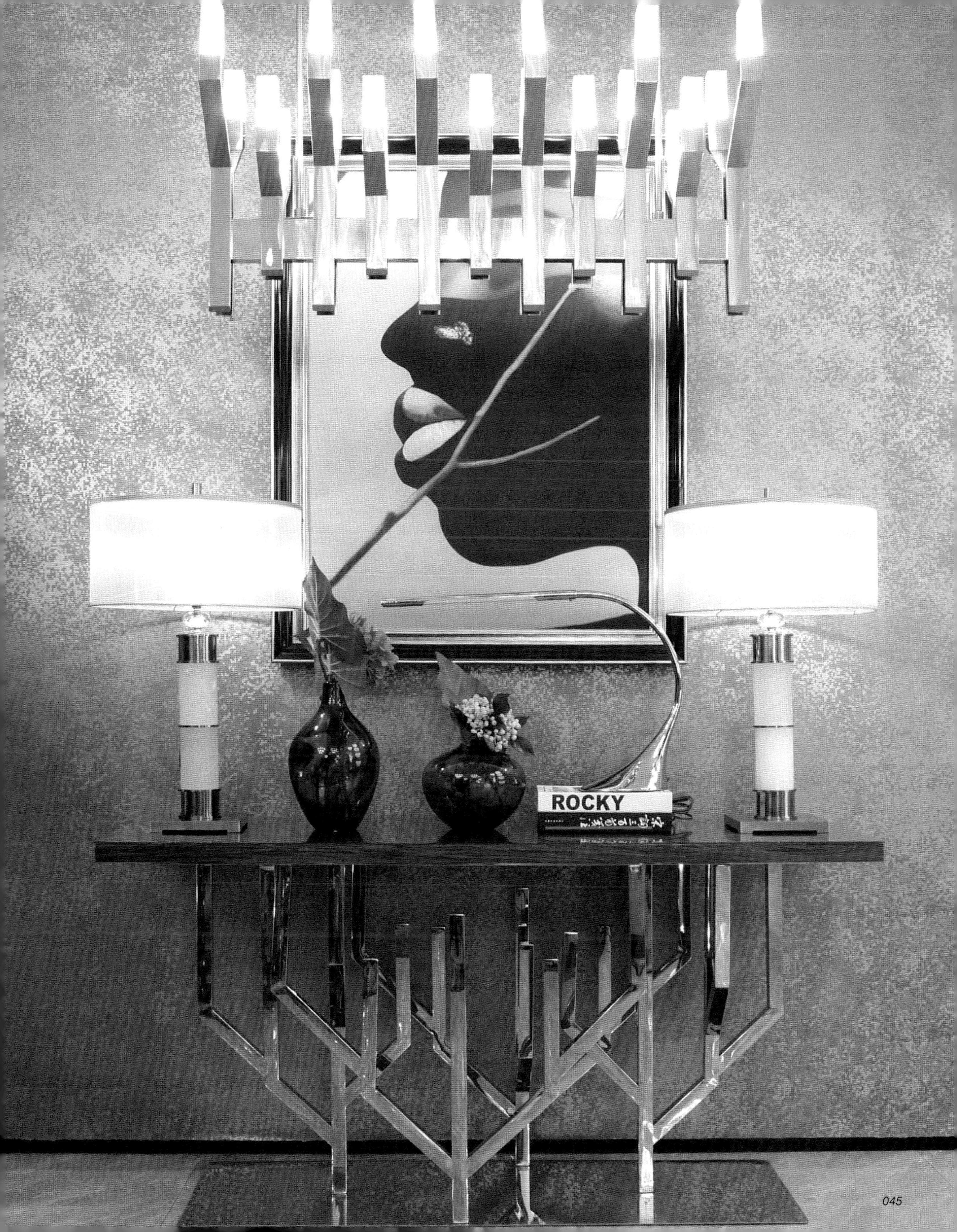
ROCKY

2.3 PRODUCT DISPLAY
单品赏析

IMAGE © OMENIA
图片来自 OMENIA 品牌

IMAGE © OMENIA

图片来自 OMENIA 品牌

Stainless Steel Vase

Streamlined stainless steel vase as a modern flower arrangement is popular in modern floriculture. In the daily modern style home furnishings, stainless steel vases and flowers are mixed to present a modern atmosphere and balanced texture creating a good decorative effect.

不锈钢花器

流线型的不锈钢花器作为插花容器是现代花艺的流行做法。在日常的现代风格家居布置中，不锈钢花器与鲜花的搭配充满现代气息，质感平衡，能起到很好的装饰作用。

01 Chess

A wise man always knows his place on the chessboard of life. And despite the hustle and bustle in our life, there is so much more we could celebrate and feel, like having a glorious victory in a chess game. Taking that spirit, this furniture collection is tailored to bring sculpture art into our daily life.

国际象棋

细腻的心思与缜密的思维，智者永远掌控着全局。人除了繁杂的生活，还有情怀，会意一局荡气回肠的凯旋，量身定做的雕塑艺术让艺术名副其实地融入生活，令栖居更艺术，让展览更磅礴。

02 Letter Vase

Modern people want their home to be more casual and relaxing. Free combination of letters series vases, will be able to meet your simple feelings for nature. Comfortable natural letters, you can always find out your favorite type, whether decorating at home or office, it is more than a ray of concise fashion charm.

字母花瓶

自然家居，现代人都希望自己的家能多一些随意与轻松，可随意组合的字母系列花瓶，就能满足你对于自然的朴素情感。舒适自然的英文字母，总有你喜爱的那一款，装点在家里或办公室，多一缕简约时尚韵味。

03 Star Series

Gorgeous space is filled with demure and beautiful atmosphere. Concise light beads create a taste of life. Life should be so free. Looking at the quiet sky and recalling the past , you appreciate fine arts like jewelry, which touches you and makes you appreciate them meticulously.

星空系列

华美的空间洋溢着娴静优美的气息，素洁轻盈的圆珠营造着生活的情趣。生活，本就该如此无拘无束。遥望着宁静的星空，追忆那似水年华，欣赏如艺术精品般的饰品，以感动化为细致品位，慢慢鉴赏。

04 Wine Set

The plain wine set and its elegant style bring us the visual enjoyment as well as the temperament improvement. You can be enthusiastic like fire, and also cherish the memorable history and stories. The wine set expresses a noble life style, which is perfectly matched with the mellow wine, directly touching you by heart.

酒具

朴素的酒架、淡雅的风采，一抹尊贵呼之欲出，这既是视觉的享受，也是气质的提升。你可以热情似火，也可以缅怀那难以割舍的历史与故事。气宇轩昂地抒写对生活品质的崇高品位，这恰好也与醇香馥郁的红酒相得益彰，直抵心灵。

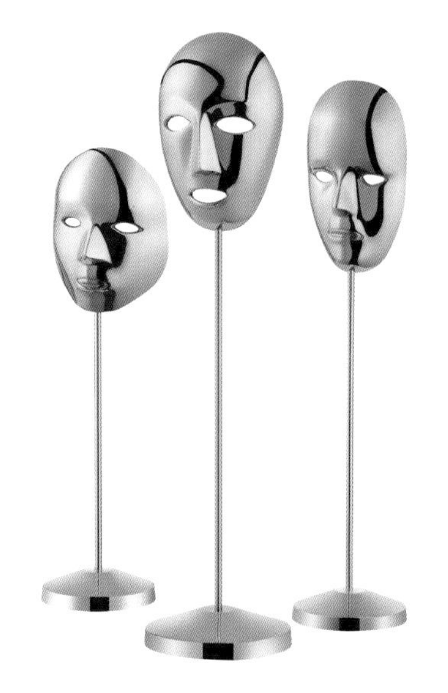

05 Sculpture Decoration

Time flies, recalling the past. After the vicissitudes, with a new attitude, it turns your wonderful memories into vivid reproduction, like magic and mottled "cross-time journey." And the difference is that all breakthroughs are condensed here.

雕塑摆件——艺术摆件

东流逝水，叶落纷纷，时光荏苒，追忆那似水年华。历经沧桑的元素，以崭新的姿态让你的美妙记忆生动再现，似幻似真，就像身临一场灿烂斑驳的"跨时空之旅"。与墨守成规不同的是，所有突破都凝聚于此。

IMAGE © OMENIA
图片来自 OMENIA 品牌

06 Round Tube Series

Simple geometric tubes are arranged in a patchwork, giving unlimited reverie. Plain shape always makes people can't help giving more glances. Ordinary things are also true — the more ordinary, the more interesting. It presents poetic space for people who love life and increases the green sense in display space.

圆管系列

朴实的几何圆管错落有致地排列，给人以无限的遐想。云淡风轻的外形，总会让人情不自禁地多看几眼，平凡的事物亦是如此，越是平凡，越是有趣，为热爱生活的人点缀出空间里的诗意。用在陈列空间里，增加空间的绿意，把自然搬进家。

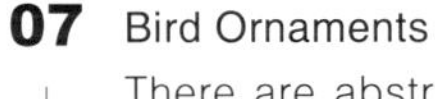

07 Bird Ornaments

There are abstract modeling, smart gesture. Stainless steel silver and champagne gold perfectly match, and enrich beauty of sculpture. The ornaments fit Omenia's concept of drawing materials from the nature. Displaying in the living room or study makes space more dynamic.

小鸟摆件

抽象的造型，灵动的姿态，不锈钢银色与香槟金色完美搭配，同时富有雕塑的美感。这也完全与欧米亚取材来源于大自然的品牌理念相得益彰。陈列于客厅、书房等空间，让空间更具生命力。

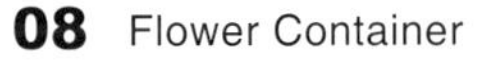

08 Flower Container

Flower container full of metal smell brings out the characteristics most vividly, making people addictive. The concept of creating everything according to law of the nature is introduced into living room, which brought us aesthetics of life. "See a world in a grain of sand, and a heaven in a wild flower." A cluster of green and touch of sweet can burst into a different kind of home style.

花器

充满金属气息的花器将以上特质表现得淋漓尽致，让人爱不释手。将道法自然的造物观引入生活与居室之中，这便是花器带给我们的生活美学。一叶一菩提、一花一世界，一簇青翠和一抹嫣然就能让家绽放出别样的风情。

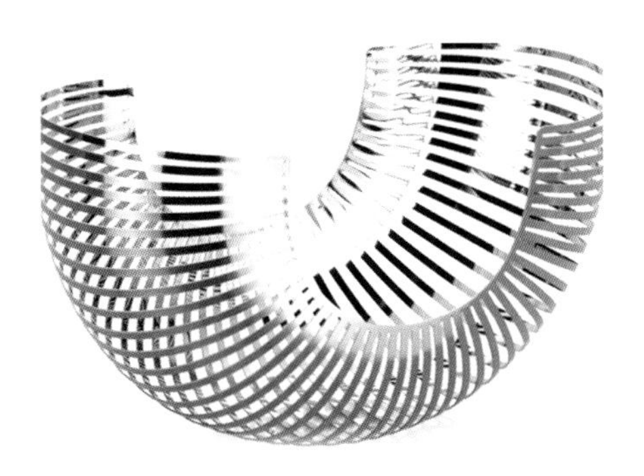

09 Fruit Plate

The entire space is filled with orderly space and the rich atmosphere of stainless steel, for the real flavor of life in them. The fruit plate's style comes from its own texture, shape, complemented by exquisite ingredients. Several blooming flowers for leisure create a different atmosphere. The pleasure mood is even more joyful.

果盘

有序的空间里，富裕的不锈钢气息充盈着整个空间，生活的真味尽在其中。果盘的风格来自它自身的质感、形状，与精美食材相辅相成，几枝绽放的花朵，为闲暇营造出不同的气氛，本就愉悦的心情，显得更加欢畅。

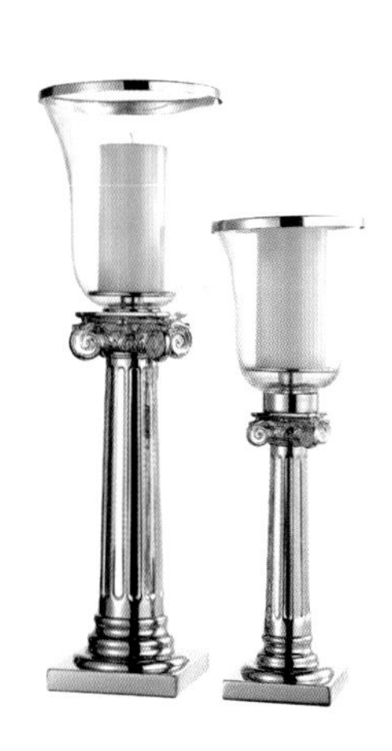

10 Candlestick

In a quiet art space, everything has been attached to the natural mysterious power. The elegant candlelight and the tender texture from stainless steel, concisely and gracefully make the whole space comfortable and perfect. Faced with the transformation of the four seasons, the room is integrating into the natural and pristine atmosphere. And the elegant life style shows the deep appreciation of life.

烛台

在纯净的艺术空间里，一切都被赋予了自然的神秘力量。典雅的烛光与温柔的不锈钢质感，将空间的舒适与完美概括成一种简洁、雅致，面对四季的变换，融入自然的质朴，精致的生活方式，演绎对生命的深层感悟。

IMAGE © OMENIA
图片来自 OMENIA 品牌

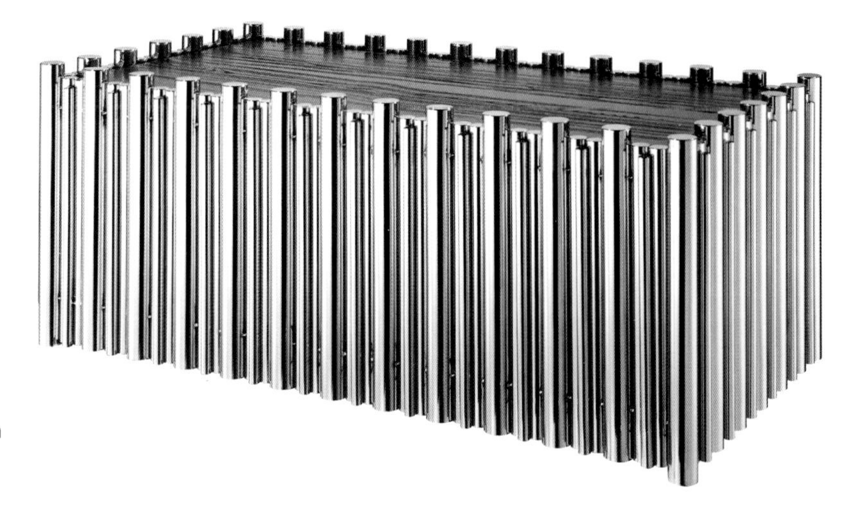

11 Hypericum Tea Table
Size I 1080mm × 560mm × 420mm

金丝桃茶几
尺寸 | 1080mm×560mm×420mm

12 Hypericum Decoration
Size I 440mm × 125mm × 520mm

金丝桃摆件
尺寸 | 440mm×125mm×520mm

13 Lotus Leaf Decoration
Size | 500mm × 210mm × 400mm
荷叶摆件
尺寸 | 500mm×210mm×400mm

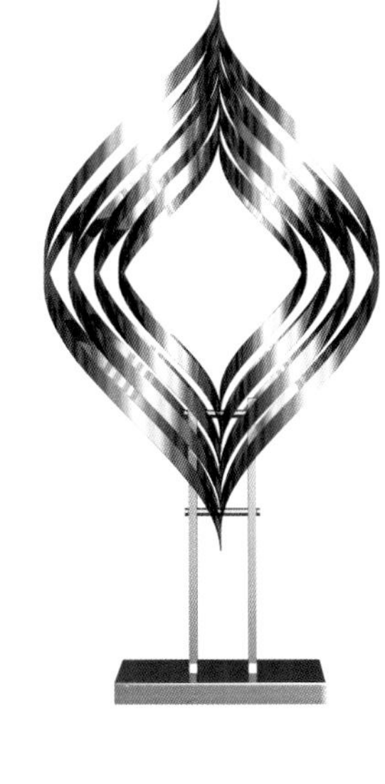

14 Sail Decoration
Size | 580mm × 200mm × 1000mm
摆件之扬帆（大）
尺寸 | 580mm×200mm×1000mm

15 Windmill
Size | 385mm × 130mm × 360mm
风车
尺寸 | 385mm×130mm×360mm

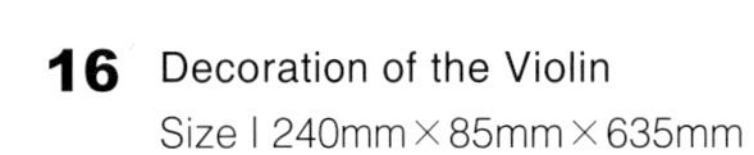

16 Decoration of the Violin
Size | 240mm × 85mm × 635mm
摆件之小提琴
尺寸 | 240mm×85mm×635mm

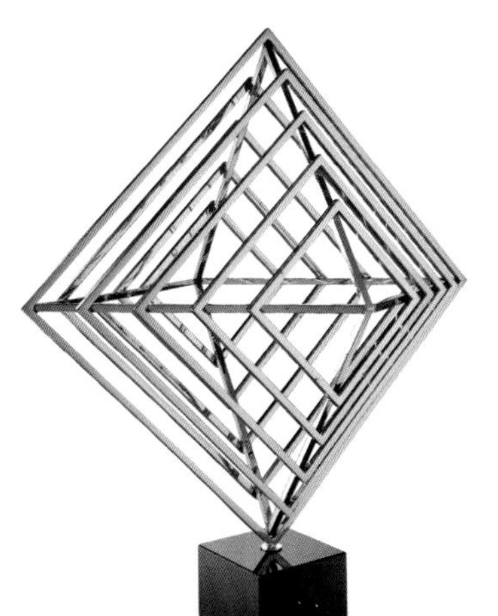

17 Champaign-Gold Rhombus Decoration
Size | 705mm × 470mm × 810mm
香槟金菱形摆件
尺寸 | 705mm×470mm×810mm

19 Ball Pile Decoration
Size | 400mm × 360mm × 640mm
球堆摆件
尺寸 | 400mm×360mm×640mm

18 Dancing Butterflies Decoration
Size | 550mm × 300mm × 830mm
舞蝶摆件
尺寸 | 550mm×300mm×830mm

20 Bauhinia Series Corner Table
Size | 600mm × 535mm × 550mm
紫金花系列之角几
尺寸 | 600mm×535mm×550mm

21 Countless Ties Decoration 1
Size | 415mm × 275mm × 530mm
千丝万缕摆件（一）
尺寸 | 415mm×275mm×530mm

22 Bauhinia Series Tea Table
Size | 1240mm × 800mm × 420mm
紫金花系列之茶几
尺寸 | 1240mm×800mm×420mm

24 Lagerstroemia Tea Table
Size I 1235mm × 605mm × 430mm

紫薇花茶几
尺寸 | 1235×605×430MM

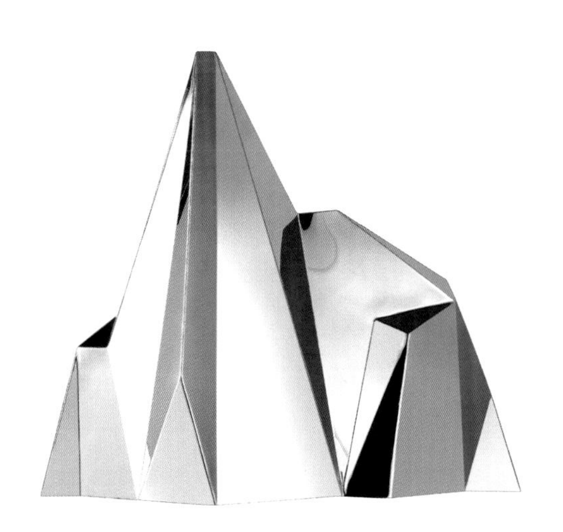

23 Rockery Decoration
Size I 700mm × 330mm × 600mm

假山摆件
尺寸 | 700×330×600MM

25 Lagerstroemia Corner Table
Size I 570mm × 570mm × 600mm

紫薇花角几
尺寸 | 570×570×600MM

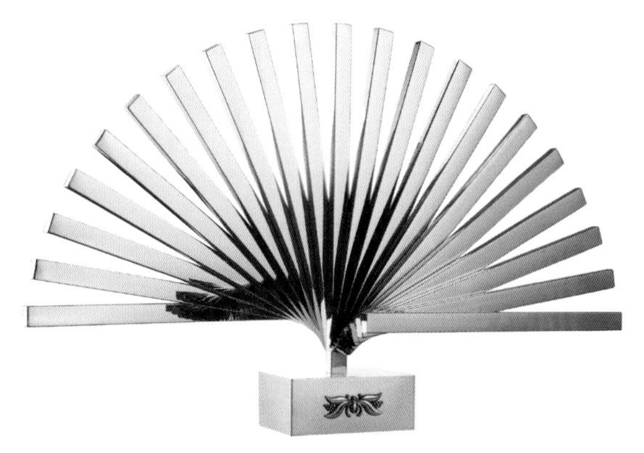

26 A Great Hawk Spreading Its Wings
Size I 960mm × 300mm × 620mm

大鹏展翅（大）
尺寸 | 960×300×620MM

27 Champaign-Gold Pentacyclic Ring Decoration
Size I 450mm × 210mm × 500mm

香槟金五环摆件
尺寸 | 450×210×500MM

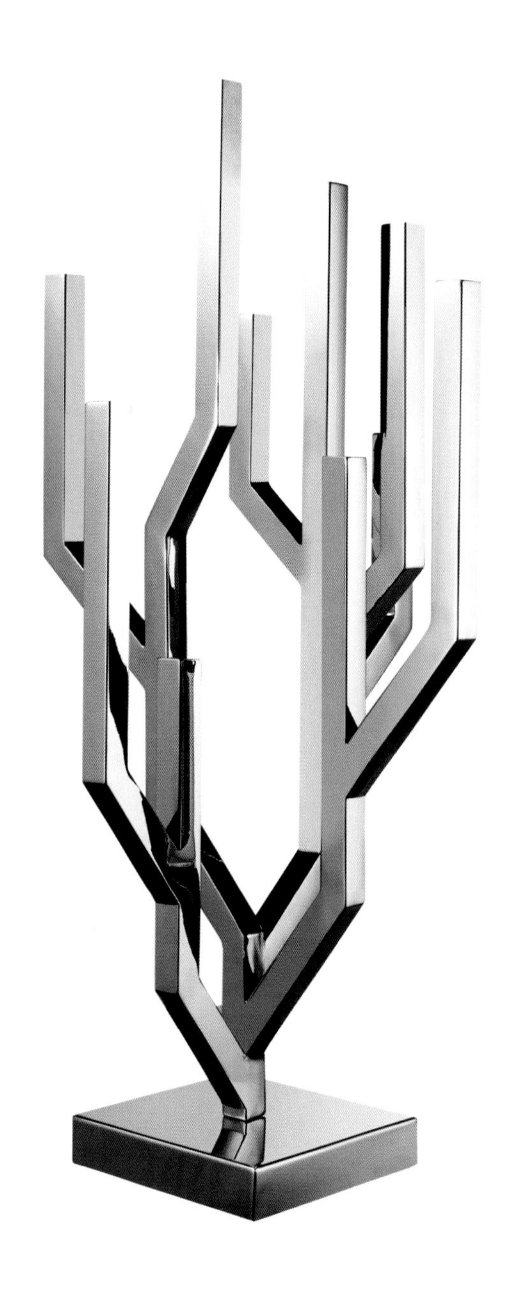

28 Yellow Chrysanthemum Decoration
Size | 335mm × 285mm × 640mm

金光菊摆件
尺寸 | 335mm×285mm×640mm

29 Branches Corner Table
Size l 500mm × 500mm × 575mm

树枝角几
尺寸 | 500mm×500mm×575mm

30 Branches Tea Table
Size l 1200mm × 600mm × 460mm

树枝茶几
尺寸 | 1200mm×600mm×460mm

31 Round Tube Corner Table
Size l 700mm × 630mm × 460mm

圆管角几
尺寸 | 700mm×630mm×460mm

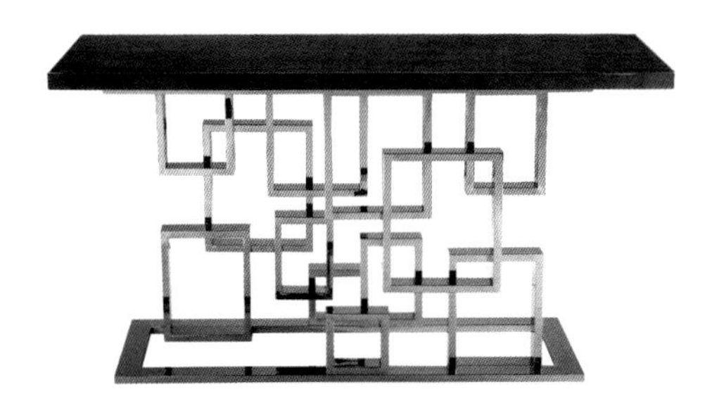

32 Cottonrose-inspired Hallway Display
Size l 1500mm × 500mm × 795mm

木芙蓉玄关
尺寸 | 1500mm×500mm×795mm

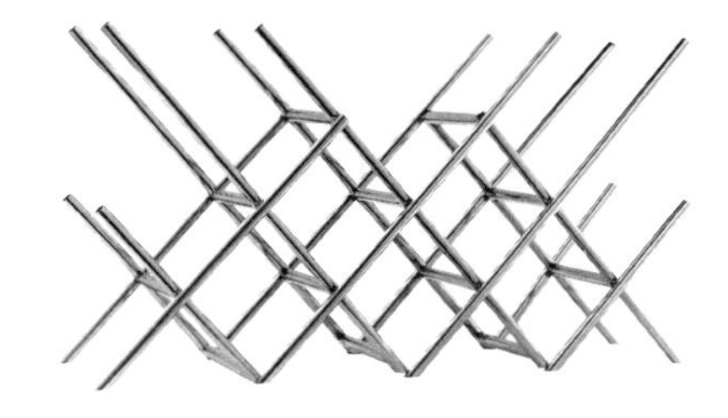

33 Woven Wine Rack
Size l 500mm × 150mm × 265mm

编织酒架（二）
尺寸 | 500mm×150mm×265mm

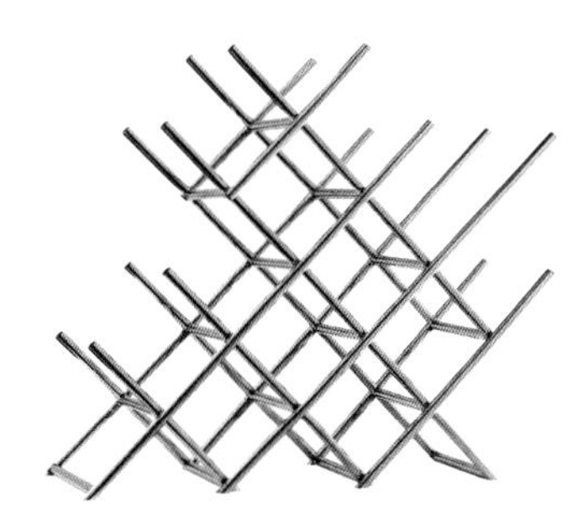

34 Woven Wine Rack
Size l 455mm × 160mm × 390mm

编织酒架
尺寸 | 455mm×160mm×390mm

IMAGE © OMENIA
图片来自 OMENIA 品牌

3

WELAND

帷澜

IMAGE © WELAND
图片来自 WELAND 品牌

IMAGE © WELAND
图片来自 WELAND 品牌

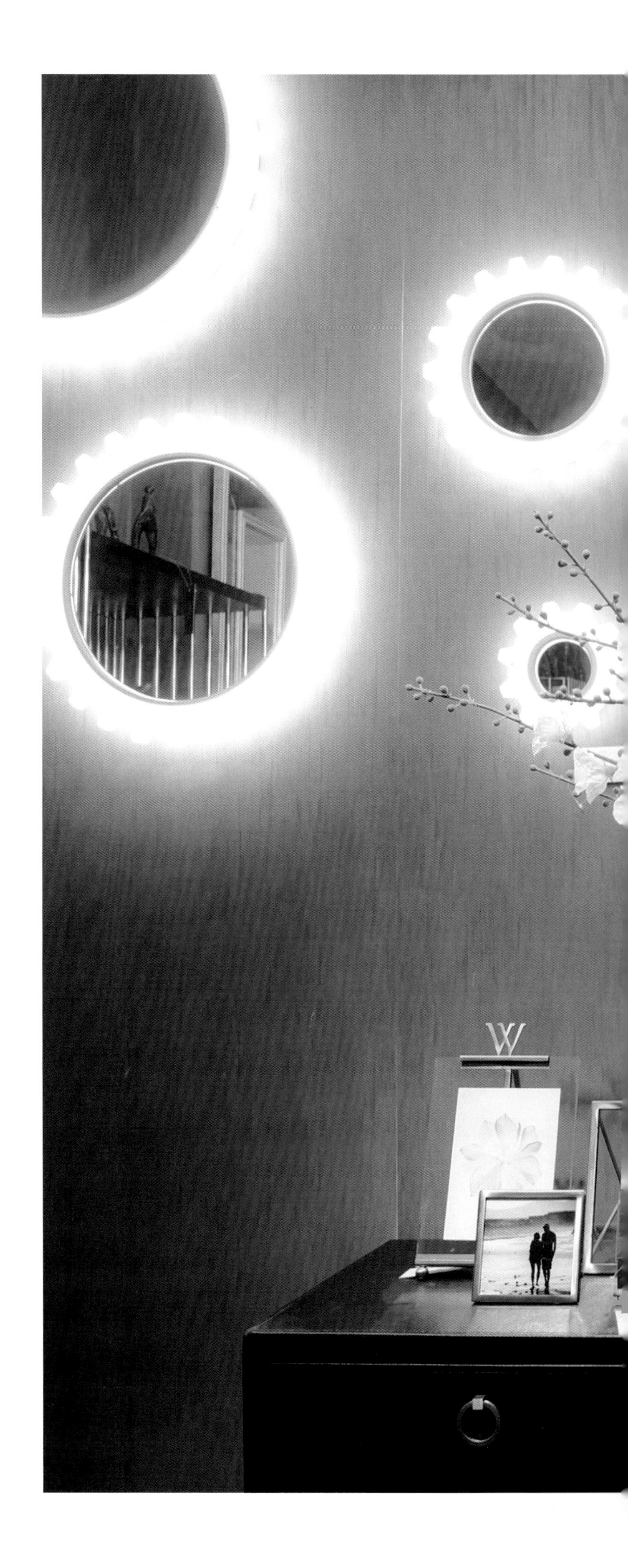

TURKS IN EUROPE
TURKS IN EUROPE
TURKS IN EUROPE

3.1 LIVING ROOM

客厅赏析

IMAGE © WELAND
图片来自 WELAND 品牌

The main colors are premium grey and bright yellow, so that the overall space does not seem boring. The room is decorated by geometric shapes, like carpet, furniture lines, pendants. Bonded with the delicate household products, the space looks concise with a sense of hierarchy.

主色调高级灰，明亮黄色，让整体空间不显得沉闷。空间采用几何图形的搭配（地毯、家具线条、挂件），结合细节精致的家居产品，简约中带有层次感。

To present a fresh and concise style, it uses white as the main color, combined with bright yellow carpet and a few orange flowers as a decoration as well as silver crystal furniture.

主色调白色，加上明黄色地毯及些许橙色花艺作为点缀，结合银色水晶系的家具，营造一种清新简约风格。

IMAGE © WELAND
图片来自 WELAND 品牌

Steady dark wood furniture and gray cloth carpet create a quiet atmosphere. The golden home accessories balance the colors and heighten atmosphere.

沉稳的深色木质家具和灰色布制地毯营造出安静的氛围，金色的家居饰品平衡了色彩，活跃了氛围。

IMAGE © WELAND
图片来自 WELAND 品牌

IMAGE © WELAND
图片来自 WELAND 品牌

Dark sofa carpet and, metal furniture and jewelry match with each other. Their texture is balanced, without a sense of dullness.

深色沙发与地毯及金属制家具和饰品相搭配，质感平衡，又无沉闷感。

IMAGE © WELAND
图片来自 WELAND 品牌

Restaurant with orange-based colors, beige wallpaper, and silver and copper furniture, has a deep sense of warmth.

餐厅以橙色为主色调，加上米白色的墙纸，配合银色和铜色家具，使空间有着浓浓的温馨感。

3.2 DINING ROOM
餐厅赏析

IMAGE © WELAND
图片来自 WELAND 品牌

3.3 BEDROOM

卧室赏析

The color of the bedroom furniture is mainly in blue, but due to the wooden materials, it does not bring out a sense of coldness. The wallpaper is in blue-green, echoed with the furniture, and the texture is also in geometric shape. The elegant color is matched with the copper jewelry, which let us feel like spring, making the whole space more vivid.

卧室家具的色调虽然偏向深蓝色，但选用了木材，也不会给人一种冰冷感。空间采用了带蓝绿色的墙纸，跟家具相呼应，墙纸的纹理也选用了几何图形。高雅的色调跟铜色饰品搭配，春天的感觉（绿叶的抱枕、绿植），让空间添加了不少生气。

IMAGE © WELAND
图片来自 WELAND 品牌

3.4 PRODUCT DISPLAY

单品赏析

IMAGE © WELAND
图片来自 WELAND 品牌

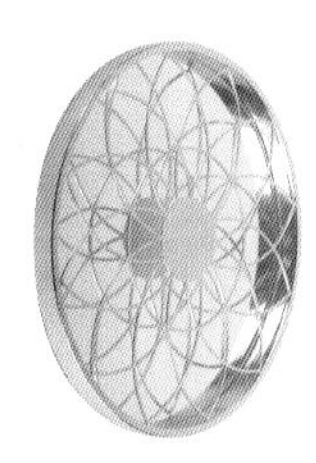

01 Hang Decoration | 60cm×3.5cm

挂饰 | 60cm×3.5cm

02 Bookend | 34cm×8cm×27cm

书立 | 34cm×8cm×27cm

03 Animal Decoration | 11cm×4cm×6cm

动物摆件 | 11cm×4cm×6cm

04 Abstract Decoration | 26cm×10cm×43cm

抽象摆件 | 26cm×10cm×43cm

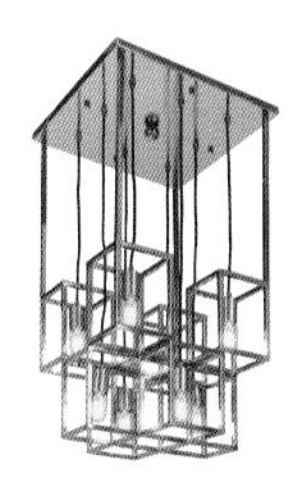

05 Chandelier | 50cm×50cm×97cm

吊灯 | 50cm×50cm×97cm

06 Storage Box | 35cm×35cm×23cm

储物盒 | 35cm×35cm×23cm

07 Clock | 22cm×22cm×22cm

钟 | 22cm×22cm×22cm

08 Candlestick | 18cm×77cm

烛台 | 18cm×77cm

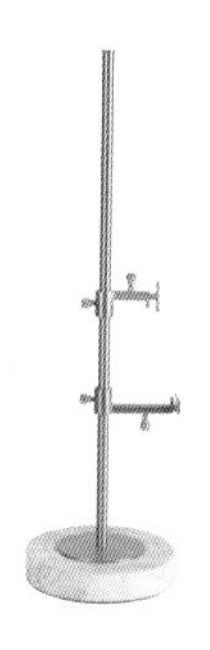

09 Photo Frame | 20.5cm×20.5cm×70.5cm

相架 | 20.5cm×20.5cm×70.5cm

10 Human Form Decoration | 43cm×15cm×25cm

人物摆件 | 43cm×15cm×25cm

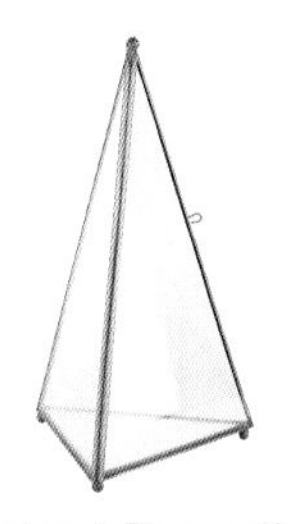

11 Abstract Decoration | 22cm×22cm×40cm

抽象摆件 | 22cm×22cm×40cm

12 Candlestick | 22cm×11cm×41cm

烛台 | 22cm×11cm×41cm

13 Photo Frame | 21cm×2cm×26cm

相架 | 21cm×2cm×26cm

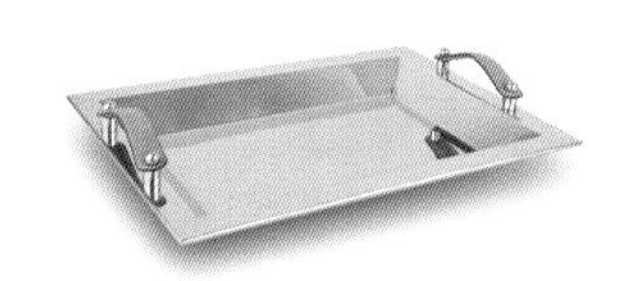

14 Tray | 40cm×25cm×6cm

托盘 | 40cm×25cm×6cm

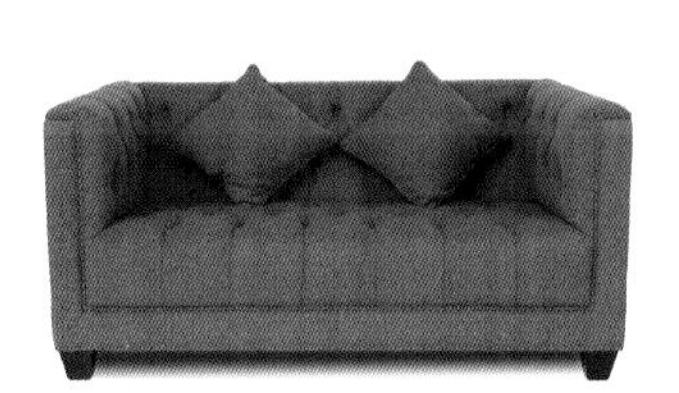

15 Double Sofa | 180cm×85cm×72cm

二人沙发 | 180cm×85cm×72cm

16 Mingshi — Candlestick | 20cm×16cm×61cm

明仕—烛台 | 20cm×16cm×61cm

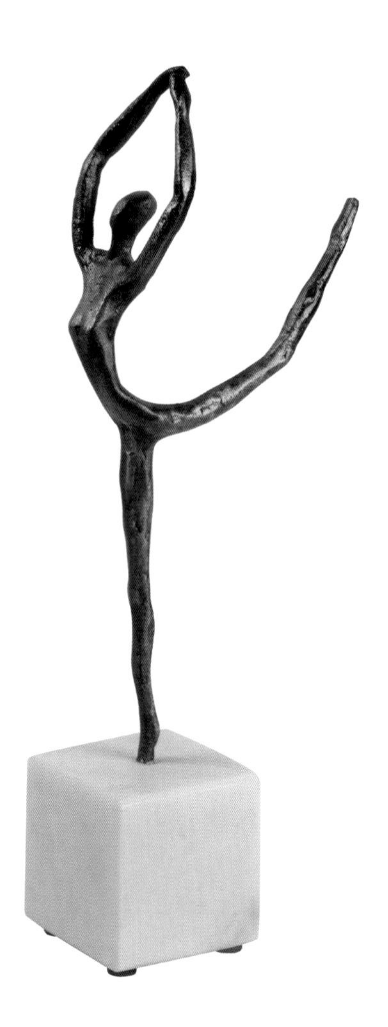

18 Table Lamp

Size I 43cm×76cm
Material I Iron & Fabric & Glass
Color I Transparent & Gold
Inspiration I Combination of Art Deco and Modern Home Furnishing
Style I Modern, American Style

台灯

尺寸 | 43cm×76cm
材质 | 铁 & 布 & 玻璃
颜色 | 透明 & 金
灵感 | 组合装饰派艺术与现代风格的家居用品
风格 | 现代、美式

17 Human Form Decoration

Size I 13cm×6cm×32cm
Material I Marble & Metallic
Color I Gold
Inspiration I Combination of Art Deco and Modern Home Furnishing
Style I Modern, Chinese Style, American Style

人物摆件

尺寸 | 13cm×6cm×32cm
材质 | 大理石 & 金属
颜色 | 金
灵感 | 组合装饰派艺术与现代风格的家居用品
风格 | 现代、中式、美式

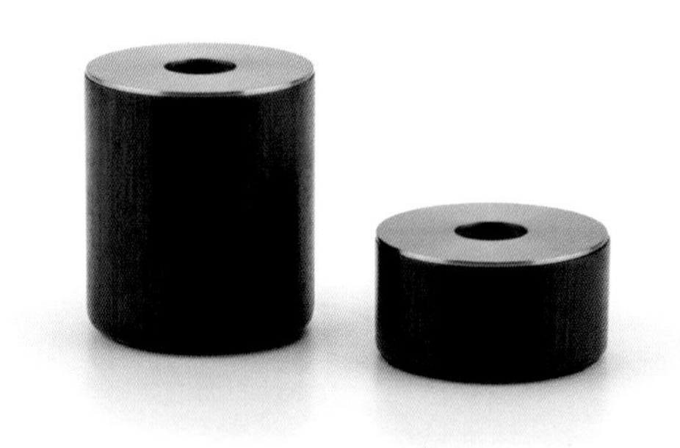

19 Candlestick

Size I 8cm×9cm / 8cm×4cm
Material I Iron & Oak Walnut
Color I Black
Inspiration I Combination of Art Deco and Modern Home Furnishing
Style I Modern, American Style

烛台

尺寸 | 8cm×9cm / 8cm×4cm
材质 | 铁 & 橡木做胡桃木
颜色 | 黑
灵感 | 组合装饰派艺术与现代风格的家居用品
风格 | 现代、美式

20 Bookend

Size I 15cm×7cm×21cm×2PCS
Material I Marble & Metal
Color I White & Gold
Inspiration I Combination of Art Deco and Modern Home Furnishing
Style I Modern, American Style

书立

尺寸 | 15cm×7cm×21cm×2PCS
材质 | 大理石 & 金属
颜色 | 白 & 金
灵感 | 组合装饰派艺术与现代风格的家居用品
风格 | 现代、美式

IMAGE © WELAND
图片来自 WELAND 品牌

21 Tray

Size I 40cm×25cm×5cm
Material I Copper & Stainless Steel
Colors I Gold & Silver
Inspiration I Combination of Art Deco and Modern Home Furnishing
Style I Modern, American Style

托盘

尺寸 | 40cm×25cm×5cm
材质 | 铜 & 不锈钢
颜色 | 金 & 银
灵感 | 组合装饰派艺术与现代风格的家居用品
风格 | 现代、美式

22 Flower Container & Candlestick

Size I 15cm×25cm
Material I Marble & Metal
Color I Transparent & Gold
Inspiration I Combination of Art Deco and Modern Home Furnishing
Style I Modern, American Style

花器 & 烛台

尺寸 | 15cm×25cm
材质 | 大理石 & 金属
颜色 | 透明 & 金
灵感 | 组合装饰派艺术与现代风格的家居用品
风格 | 现代、美式

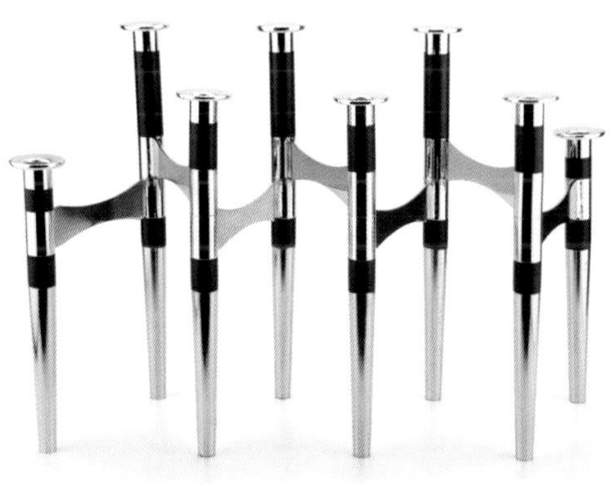

24 Candlestick

Size I 120cm×5cm×37cm
Material I Iron
Color I Gold / Black
Inspiration I Combination of Art Deco and Modern Home Furnishing
Style I Modern, American Style

烛台

尺寸 | 120cm×5cm×37cm
材质 | 铁
颜色 | 金 / 黑
灵感 | 组合装饰派艺术与现代风格的家居用品
风格 | 现代、美式

23 Snack Shelves

Size I 30cm×15cm×35cm
Material I Iron & Glass
Color I Gold
Inspiration I Combination of Art Deco and Modern Home Furnishing
Style I Modern

点心架

尺寸 | 30cm×15cm×35cm
材质 | 铁 & 玻璃
颜色 | 金
灵感 | 组合装饰派艺术与现代风格的家居用品
风格 | 现代

25 Magazine Rack

Size I 56cm×15cm×47cm
Material I Iron & Leather
Colors I Orange & Gold
Inspiration I Combination of Art Deco and Modern Home Furnishing
Style I Modern, Neo-Chinese Style

书报架

尺寸 | 56cm×15cm×47cm
材质 | 铁 & 皮
颜色 | 橙 & 金
灵感 | 组合装饰派艺术与现代风格的家居用品
风格 | 现代、新中式

ORIENTAL NEW ARISTOCRACY

东方新贵

Elegant life
雅致生活

China is a poctic country, also an important source of oriental civilization.

From the tranquil leisure of "For I pick chrysanthemums under the eastern hedge, and far away to the south I can see the mountains," to the simple warmth of "There's a gleam of green in an old bottle. There's a stir of red in the quiet stove," from the romantic elegance of "Poetry and self-entertainment, fish are kept for viewing, cook tea and play with crane," to the peace of mind of "Detached view of the secular world, calm and comfortable in mind", ancient literati outline a quiet indifferent and harmonious cozy paradise for us.

With the evolution of history, now we have entered a highly developed industrial and commercial society. In a generally more impulsive era, people are always walking anxiously. In the modern bustling city filled with reinforced concrete, more people want to slow down to have an artistic and poetic life. "Look at the three thousand years of history, all people seek for whether fame, or wealth. But when you see through life, the ultimate quest is only plain life."

In the new era, the intellectuals are mostly highly educated, with a profound cultural background and knowledge connotation. They are playing an important role in various fields of society, with the elegant name "modern Oriental noble". These people pursue the sentiment, and their demand for life is beyond the material itself. They also want to enjoy themselves spiritually, so they desire an elegant and lively living place, surrounded by the cultural atmosphere. The intellectuals put more emphasis on the comfort and functionality, but not magnificent in deliberate. This tendency turns out that the Neo-Chinese life style is accepted by its gentle and elegant sentiment nowadays when the minimalism is popular. The honorable culture and the peaceful life style gradually return to our society, representing the recall for the rural China times. It is also a spiritual yearning of the idyllic life as well as the grand respect for the tranquil soul.

Elegant life, the pursuit of a refined attitude to life, is a blend of ancient and modern neo-classical way of life. It is from small to understand big things, and it cares more about consciousness than ceremony and more spirit than material. The Chinese elegant life will be out of the question without deep culture heritage and profound ideological content. With small bridge and water, pleasant music, listening to Kunqu Opera in front of the pavilion, travel the Taihu Lake in the snow with friends, together with a cup of tea, a pool of lotus, elegant life is so simple and easy to get. When autumn leaves fall in the rain, playing chess game, or the zither, or drawing, to find a relaxing place for the body in the elegant coziness, in fact, is a perfect choice for healthy temperament.

Zhou Zuoren also said: "We look at the sunset, watching the autumn river and the flowers, listening to the rain, smelling and drinking wine not to quench thirst, eating the snacks, though useless decorate, but the more refined the better." This is Chinese-style elegant life.

中国是一个诗的国度，也是东方文明的重要发源地。

从"采菊东篱下，悠然见南山"的恬淡闲适，到"绿蚁新醅酒，红泥小火炉"的质朴温暖；从"洗砚鱼吞墨，烹茶鹤避烟"的风流雅致，到"笑看风轻云淡，闲听花静鸟喧"的怡然自得，历代文人为我们勾勒出了一个淡泊宁静、和谐惬意的世外桃源。

随着历史车轮的演进，如今我们已进入一个高度发达的工商社会。在一个人心普遍比较浮躁的时代，人们总是步履纷杂、行色匆匆。在钢筋混凝土弥漫的现代繁华都市里，多少人希望自己能慢下来，过一种艺术的、诗意的生活。"三千年读史，不外功名利禄。九万里悟道，终归诗酒田园。"

新时代下的高知群体，大多受过高等教育，具有深厚的文化背景和知识内涵，在社会各领域担当重要角色，有着"东方新贵"的雅称。他们追求情调，对生活的需求超越物质本身，推崇精神领域的求索和享受，对居室空间要求雅致活泼有生机，文化气息浓厚，强调舒适和功能性，不刻意富丽堂皇。这也直接导致在"极简主义"当道的今天，具有"禅味"的新中式生活正以其娴静雅致的生活情调、淳厚尊贵的文化脉系与淡泊空灵的悠远意境逐渐回归，意味着对"乡土中国"的一种历史追忆，是对农耕时代田园牧歌生活的一种精神向往，也是对质朴恬淡心灵境界的一种崇尚回归。

雅致式生活追求一种精致的生活态度，是一种融汇古今的新古典生活方式，它以物入微，以式入道，它重仪式更重意识，求物质更求精神，如果没有深厚的文化底蕴与博大精深的思想内涵，那么中式雅致生活也就无从谈起。小桥流水，丝竹于耳，亭前听昆曲，雪中泛太湖，啸聚同好，一席茶，一池荷，熏香迟暮，花馔青灯，雅致生活如此简单易得。红叶飘雨之际，或棋盘对弈，或抚琴一曲，或取丹青于案头，在优雅的惬意中给身心一个放松的归所，实为调养性情的不二之选。

周作人先生也说："我们看夕阳，看秋河，看花，听雨，闻香，喝不求解渴的酒，吃不求饱的点心，都是生活上必要的——虽然是无用的装点，而且是愈精炼愈好。"这就是——中国式雅致生活。

Elegant and unique,
beautiful and
unconventional
——高雅别致，美好而不落俗套

IMAGE © EASE WORKSHOP
图片来自自在工坊品牌

IMAGE © EASE WORKSHOP
图片来自自在工坊品牌

Chinese-style Elegant Charm
中式雅致风韵

The Neo-Chinese Style is born in China's traditional culture revival period, accompanied by enhanced national strength and national consciousness. In the exploration of Chinese design community at the beginning of the local consciousness, mature new generation of design teams and consumer market nurture subtle beautiful Neo-Chinese Style. In the era of Chinese culture sweeping the world, soft Chinese elements and modern materials, the Ming and Qing Dynasties furniture, window lattices, fabric bed products bring out the best in each other, reproducing the scene of the exquisite pieces.

Concise aesthetic of classical Chinese charm, graceful character for thousands of years, transforms into a new look. The introverted and calm ancient China as the source, integrate into elements of fashion and pragmatism expression techniques. Ancient and modern, this infusion between Eastern and Western cultures, would definitely generate something incredible as a result, which also adds a distinctive elegance to the renovated Chinese-style, while maintaining the calmness from the traditional Chinese heritage.

新中式风格诞生于中国传统文化复兴时期，伴随着国力增强、民族意识复苏，在探寻中国设计界的本土意识之初，逐渐成熟的新一代设计队伍和消费市场孕育出含蓄秀美的新中式风格。在中国文化风靡全球的时代，中式元素与现代材质的巧妙兼柔，明清家具、窗棂、布艺床品相互辉映，再现了移步变景的精妙小品。

凝练唯美的中国古典情韵，数千年的婉约风骨，以崭新的面貌蜕变舒展。以内敛沉稳的古意中国为源头，融入时尚元素与实用主义的表现手法，古老与现代、东方与西方，两种文化相得益彰而又水乳交融，格调品位毋庸置疑，传统文化那淡然悠远的人文气韵平添一份无双的风雅。

Performance Techniques

The Neo-Chinese style pays attention to regularity and symmetry, balancing the concept of yin-yang to reconcile the indoor ecology. The use of natural decorative materials and "gold, wood, water, fire, soil," such combination of Five Elements creates a Zen-style rational and tranquil environment.

The space is often decorated with concise and clean straight lines, sometimes with panel-type furniture influenced by Western industrial design. Linear decoration in the use of space, not only reflects the pursuit of modern simple living requirements, but also does it meet the requirement of the Chinese furniture to pursue introverted, simple design style.

Decorative Space

The Neo-decoration is very particular about creating special levels. According to the capacity and privacy requirement, different functional spaces are often divided by cased openings or antique curio shelves; to keep certain space out of sight, screens and lattice window are often applied. With those approaches, unit-style residential house shows the beauty of Chinese-style home, especially in the smaller housing, which often results in "walk in scene" decorative effect. In the decorative details of respect for natural appeal, flowers, birds, fish and others, uncompromising attention to detail, full of change, all of which fully embodies the spirit of traditional Chinese aesthetics.

Modeling

Decorative space uses more concise and tough straight lines. The use of linear decoration in space, not only reflects the pursuit of modern simple living requirements, but also does it meet the requirement of the Chinese furniture to pursue introverted, simple design style, which makes the Neo-Chinese style more practical and modern.

Color

The Neo-Chinese style furniture is mostly dark-based color, matching with the wall color: firstly, the tone is based in black, white, gray of Suzhou gardens and Beijing houses; secondly, on the basis of the black, white and gray colors, red, yellow, blue, green colors and so on of the royal residence are as a local color.

Furniture

The Neo-Chinese style could be classical furniture only, or the combination of modern and classical furniture. The Ming and Qing Dynasty furniture is the representative of the Chinese classical furniture. In the Neo-Chinese style furniture, accessories take the concise-line style of Ming Dynasty style furniture as the principal thing.

Decorative Element

Silk, Yarn, Fabric, Cloth, Wallpapers, Glass, Antique Tiles, Marble, Calligraphy and Painting, Plaque, Hanging Screen, Chinaware, Antique, Earthenware, Screen, Personal Galleries, Round-backed Armchair, Carved Wood Window, Bonsai, Hydroponics, as well as a certain meaning of Chinese classical items, exquisite porcelain, meaningful decorative painting, perfectly introduce the passion collision of history and modern, classic and fashion.

表现手法

新中式风格讲究纲常，讲究对称，以阴阳平衡概念调和室内生态。选用天然的装饰材料，运用“金、木、水、火、土”五种元素的组合规律来营造禅宗式的理性和宁静环境。

空间装饰采用简洁、硬朗的直线条，有时还会采用具有西方工业设计色彩的板式家具，搭配中式风格来使用，直线装饰在空间中的使用，不仅反映出现代人追求简单生活的居住要求，更迎合了中式家具追求内敛、质朴的设计风格。

装饰空间

新中式风格非常讲究空间的层次感，依据住宅使用人数和私密程度的不同，需要做出分隔的功能性空间，一般采用“哑口”或简约化的“博古架”来区分；在需要隔绝视线的地方，则使用中式的屏风或窗棂，通过这种新的分隔方式，单元式住宅就展现出中式家居的层次之美，尤其是在面积较小的住宅中，往往可以达到“移步变景”的装饰效果。而在装饰细节上崇尚自然情趣，花鸟、鱼虫等精雕细琢，富于变化，充分体现出中国传统美学精神。

造型

空间装饰多采用简洁硬朗的直线条。直线装饰在空间中的使用，不仅反映出现代人追求简单生活的居住要求，更迎合了中式家具追求内敛、质朴的设计风格，使“新中式”更加实用、更富现代感。

色彩

新中式风格的家具多以深色为主，墙面色彩搭配：一是以苏州园林和京城民宅的黑、白、灰色为基调；二是在黑、白、灰基础上以皇家住宅的红、黄、蓝、绿等作为局部色彩。

家具

新中式风格的家具可为古典家具，或现代家具与古典家具相结合。中国古典家具以明清家具为代表，在新中式风格家具配饰上多以线条简练的明式家具为主。

装饰元素

丝、纱、织物、布艺、壁纸、玻璃、仿古瓷砖、大理石、字画、匾幅、挂屏、瓷器、古玩、陶艺、屏风、博古架、圈椅、木雕窗花、盆景、水培植物以及具有一定含义的中式古典物品等，精美的瓷器、寓意深刻的装饰画等，完美演绎历史与现代、古典与时尚的激情碰撞。

4

EASE WORKSHOP

自在工坊

IMAGE © EASE WORKSHOP
图片来自自在工坊品牌

IMAGE © EASE WORKSHOP
图片来自自在工坊品牌

Weather changing with time, natural energy floating through the land, the beauty of the material, superb craftsmanship, only when the four of them are considered together, something great can be created.

天有时，地有气，材有美，

工有巧，合此四者，然后可以为良。

4.1 LIVING ROOM
客厅赏析

In the Chinese-style interior design, we should pay attention to light and dark colors arrangement. Placing some light-colored decorations in the dark furniture is a good match.

在中式室内设计中，要注意浅色系和深色系的搭配。在深色家具上放置一些浅色的装饰品是不错的搭配方式。

IMAGE © EASE WORKSHOP
图片来自自在工坊品牌

IMAGE © EASE WORKSHOP
图片来自自在工坊品牌

In the living room, lines of wooden sofa are simple, and there is sense of texture rather cumbersome. The use of light-colored pillow and cushion with dark wood sofa, is neutralizing the serious sense.

在客厅中，木质沙发线条干脆，有质感而无累赘感。采用浅色的抱枕和坐垫搭配深色系木质沙发，又中和了严肃感。

IMAGE © EASE WORKSHOP
图片来自自在工坊品牌

The moderate-sized ornaments in the center become a visual focus. Two book shelves are in symmetrical distribution, giving stability to the layout.

体积适中的装饰品摆放在正中间，成为视觉中心。两个书架对称分布，布局有稳定感。

IMAGE © EASE WORKSHOP
图片来自自在工坊品牌

Hangzhou Wangxing Ji Fan Exhibition Hall

杭州王星记扇业展厅

In the more open space, you can use the plain beige or white walls decorated with a large traditional Chinese painting works, but its color must be coordinated to other furniture.

在较为开阔的空间里，可以使用在素净的米色或白色墙面装饰大幅的国画作品，但色彩必须和其他家具协调。

Chinese interior design style is usually calm. So in the living room and other space, you can use a number of pillows with lively colors and rich patterns to heighten the atmosphere.

中式室内设计风格通常较为沉稳，在客厅等空间，可利用一些花纹丰富颜色较为鲜艳的抱枕来活跃气氛。

IMAGE © EASE WORKSHOP
图片来自自在工坊品牌

IMAGE © EASE WORKSHOP
图片来自自在工坊品牌

In the linear Treasure Pavilion with rounded but curve vases and other ornaments, you can break a sense of seriousness brought by the straight line.

在直线形的百宝阁里摆放线条圆润有弧度的花瓶等装饰品，可以打破直线带来的严肃感。

4.2 STUDY ROOM

书房赏析

IMAGE © EASE WORKSHOP
图片来自自在工坊品牌

The traditional tough-lined wooden furniture with light-colored wall is a common mix of Chinese-style home furnishing.

以线条硬朗的传统木质家具搭配浅色墙面是中式家居的常见搭配。

IMAGE © EASE WORKSHOP
图片来自自在工坊品牌

Chinese style furniture with the full realization of ancient charm, has brought a very serious, elegant feeling. The whole setting allows that we can be truly exposed to the mood of ancient literati and fully aware of the majestic traditional culture.

中式家具搭配完全实现了一种古代的风韵，带来了一种非常严肃、高雅的感觉，整个设置让大家能够真正地置身于古代文人墨客的心境之中，充分地感知传统文化的磅礴。

IMAGE © EASE WORKSHOP
图片来自自在工坊品牌

4.3 BEDROOM
卧室赏析

The use of the same series of Neo-Chinese style furniture helps the archaistic home furnishing blend in with modern elements, which avoids producing a very strong sense of ancient elegance but full of modern Chinese fashion personality style.

空间中使用同一系列的新中式家具让拥有着古风的家居，融合到现代化的元素当中，以至于没有产生非常浓烈的古老典雅的感觉，反而充满着时尚中国现代化的个性风格。

IMAGE © EASE WORKSHOP
图片来自自在工坊品牌

White space is an important element in Chinese painting. It also plays a key role in Chinese design. The clean walls, dark furniture and floors create a low-key and quiet atmosphere. Wall is not linked to decoration, but bright and low-key.

留白是国画中的重要元素，在中式室内设计中，留白也是重要的组成部分。素净的墙面与深色家具和地板营造出低调幽静的氛围。墙面不挂装饰，敞亮低调。

4.4 PRODUCT DISPLAY

单品赏析

01 Refreshing Breeze | Leisure Chair

Size | 650mm×560mm×860mm
Comfortable breeze, and ethereal quiet.
"Emptiness" means inclusiveness and greater possibilities. Works with clear water-like acrylic material show the concept of "emptiness", enabling the chair maintain this calm sprit under different circumstances.

清风 | 休闲椅

尺寸 | 650mm×560mm×860mm
自在清风，空灵静谧。
"空"意味着包容及更大的可能性。作品以清水般剔透的亚克力材质表现"空"的概念，使得椅子在各种环境氛围下都能体现静谧气息。

IMAGE © EASE WORKSHOP
图片来自自在工坊品牌

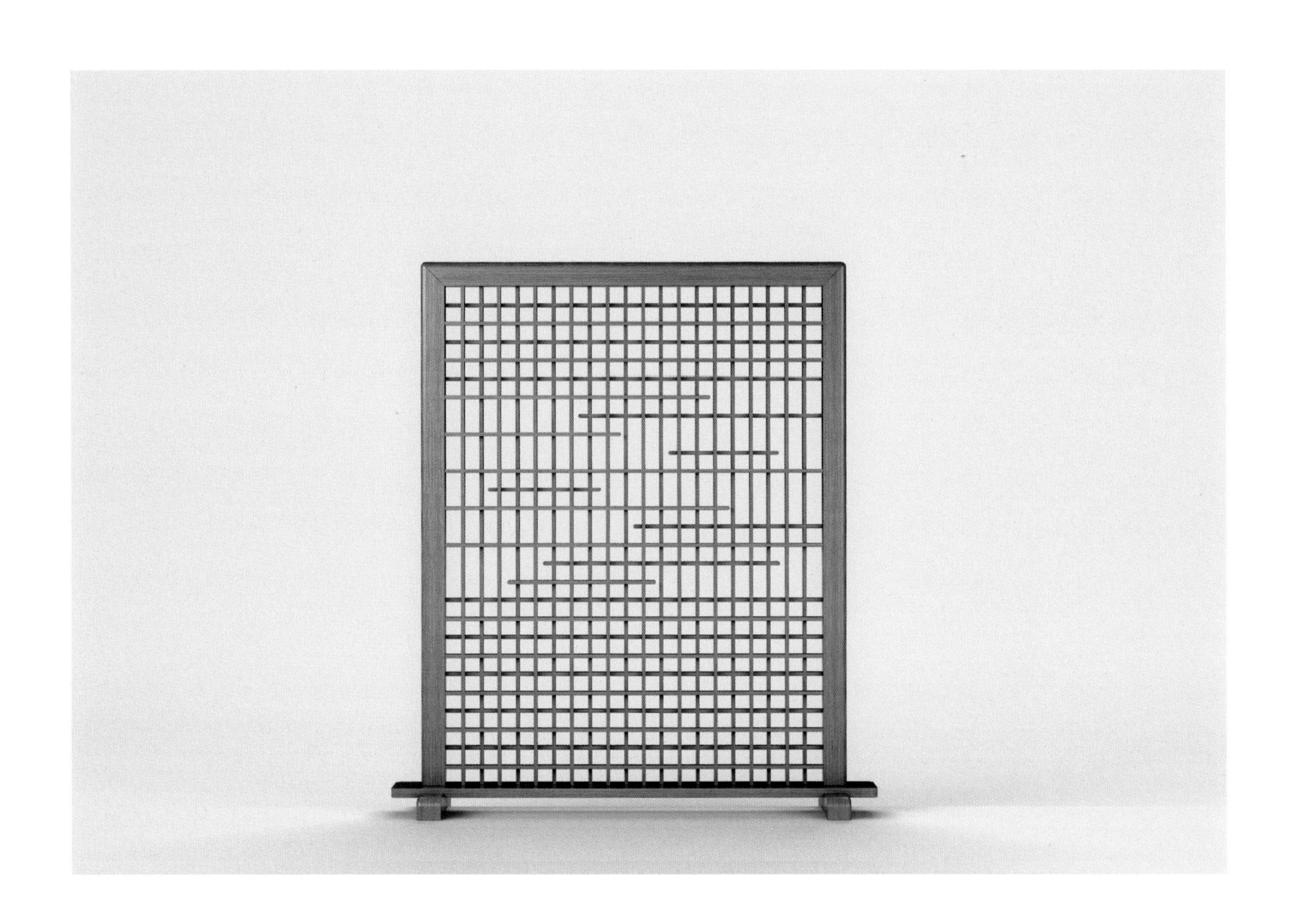

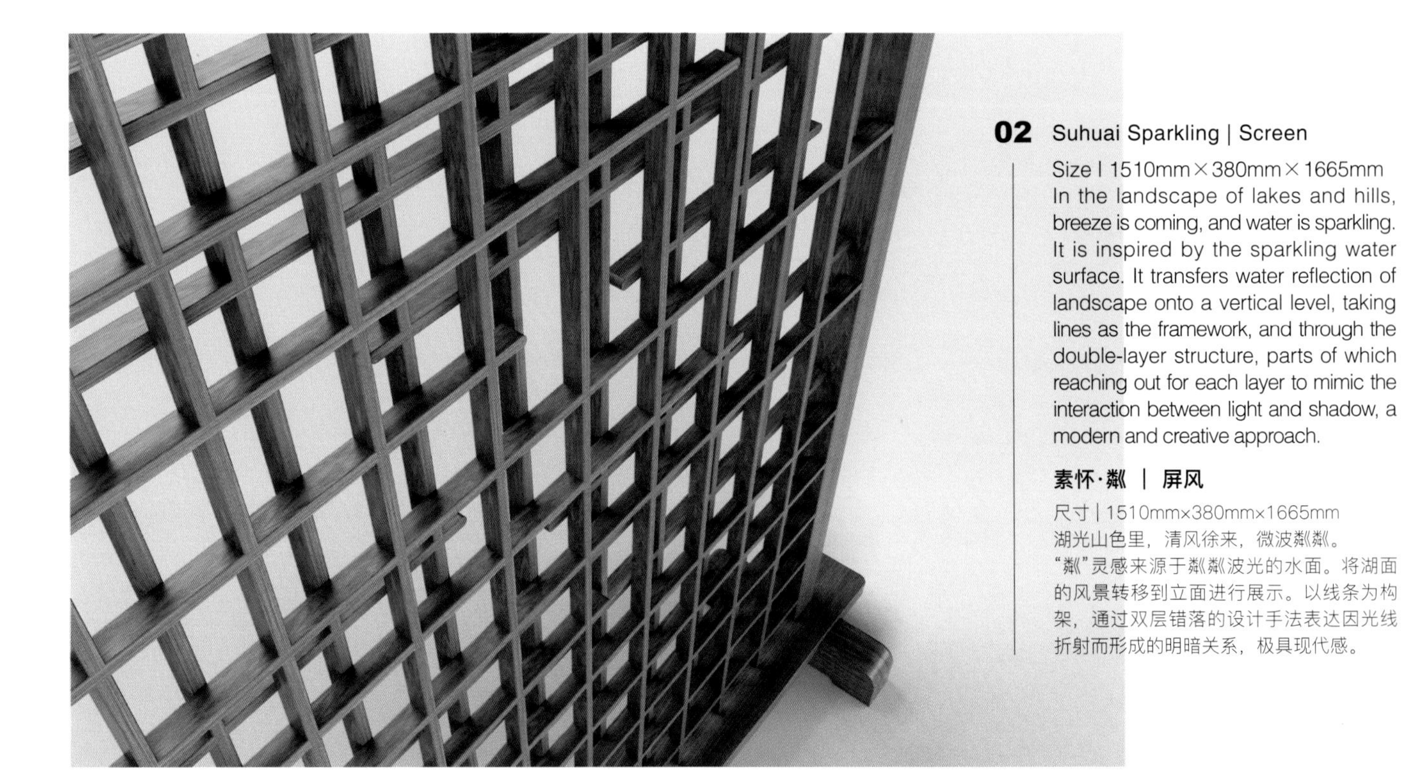

02 Suhuai Sparkling | Screen

Size | 1510mm × 380mm × 1665mm
In the landscape of lakes and hills, breeze is coming, and water is sparkling. It is inspired by the sparkling water surface. It transfers water reflection of landscape onto a vertical level, taking lines as the framework, and through the double-layer structure, parts of which reaching out for each layer to mimic the interaction between light and shadow, a modern and creative approach.

素怀·粼 | 屏风

尺寸 | 1510mm×380mm×1665mm
湖光山色里，清风徐来，微波粼粼。
"粼"灵感来源于粼粼波光的水面。将湖面的风景转移到立面进行展示。以线条为构架，通过双层错落的设计手法表达因光线折射而形成的明暗关系，极具现代感。

IMAGE © EASE WORKSHOP
图片来自自在工坊品牌

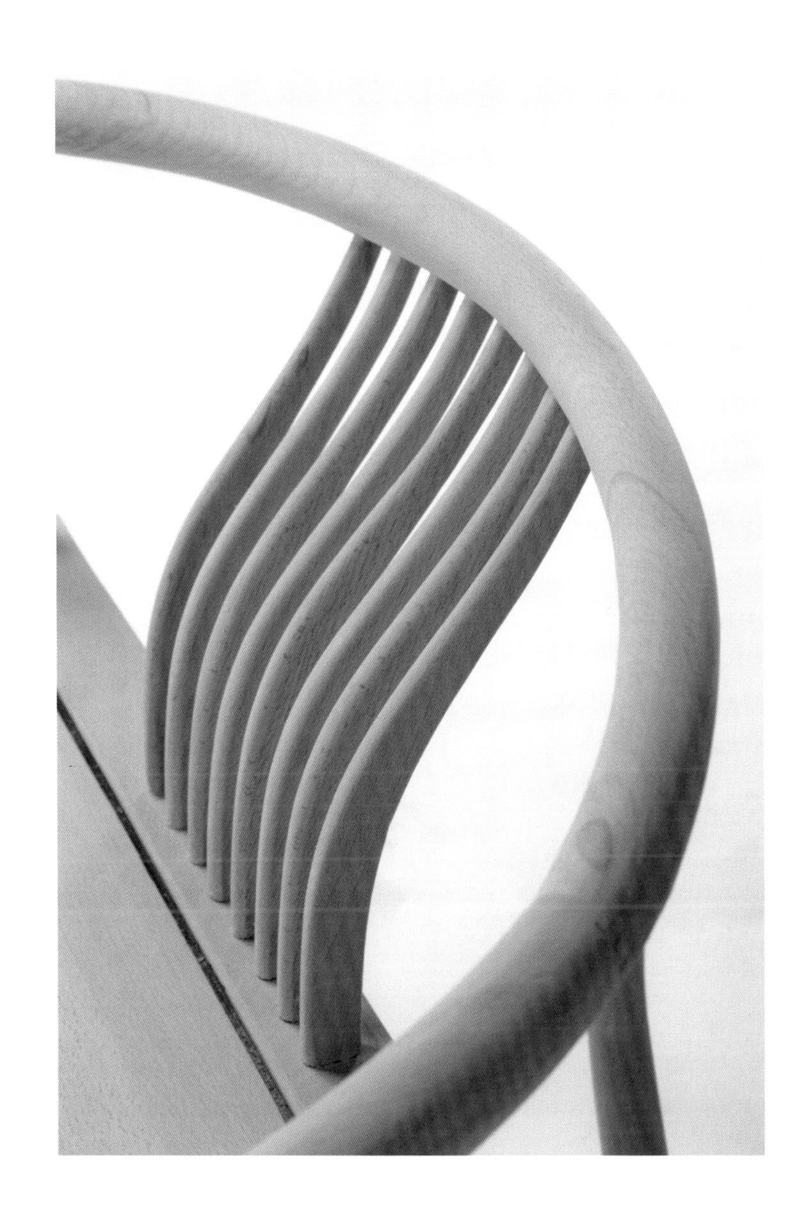

03 Suhuai | Free Tune

Size I 680mm×593mm×815mm
Blowing the seven-string instrument, letting the music echo with running spring. Utilizing wood, cloth and acrylic as materials, the Free Tune expresses the flexibility and inclusiveness of liquid sprit and also the material colors transcend a gentle feeling.

素怀·曲漫

尺寸 | 680mm×593mm×815mm
拂动七弦，鸣泉听水。
曲漫以微妙的变化，表现水流的灵动与丰富的包容性。结合木、布艺、亚克力三种材质，着重体现三者的温和之感。

04 Suhuai Slow Life | Zen Chair

Size I 950mm×830mm×620mm
The so-called Zen, assembles breeze and diffuses fragrance.
Designers believe that the pace of modern life is compact, which is so in need of silence and introspection time. It will re-design Zen chair to break its inherent image, and change its form and proportion to better integrate into the modern home furnishing life.

素怀·慢生 | 禅椅

尺寸 | 950mm×830mm×620mm
所谓禅，折聚清风，漫惜余香。
设计师认为现代生活节奏紧凑，需要沉寂和自省的时间。故将禅椅重新进行设计，打破其固有形象，改变其形式、比例，使其更好地融入现代家居生活之中。

IMAGE © EASE WORKSHOP
图片来自自在工坊品牌

05 Gold and Jade · Bright | Garden Stool

Size l 44mm×380mm×470mm
Shadow changes from time to time.
Appearance is sourced from the Chinese classical furniture. Garden stool frame is cut by laser, and lighting could be installed inside. "Leaves and flowers" mapped to the entire space creating a romantic art for life.

金玉·有明 | 鼓凳

尺寸 | 440mm×380mm×470mm
落花有时，余影婆娑。
外观源于中国古典家具中的鼓凳，通过金属激光切割制作而成，内部可放置灯，"落叶与花"映射到整个空间，为生活融入浪漫的艺术气息。

IMAGE © EASE WORKSHOP
图片来自自在工坊品牌

06 Gold and Jade · Frame | Leisure Chairs

Size | 630mm×590mm×840mm
It joins curved and straight, virtual and real, dynamic and static elements together.
The chair is designed to explore the collision between modern metal and oriental aesthetics. Through the forging process, the combination of curved and straight lines, the contrast of the thickness and the leather material, the chair presents a modern oriental charm.

金玉·局 | 休闲椅

尺寸 | 630mm×590mm×840mm
曲直相接，虚实相生，语默动静体自然。
椅子的设计意在探索现代金属与东方美学的碰撞，通过锻造工艺，曲直结合，粗细的对比，配合皮革材料，使椅子呈现出一种现代时尚的东方神韵。

IMAGE © EASE WORKSHOP
图片来自自在工坊品牌

5

CHUN ZAI DONG FANG

春在东方

IMAGE © CZDF
图片来自春在东方品牌

5.1 LIVING ROOM
客厅赏析

IMAGE © CZDF
图片来自春在东方品牌

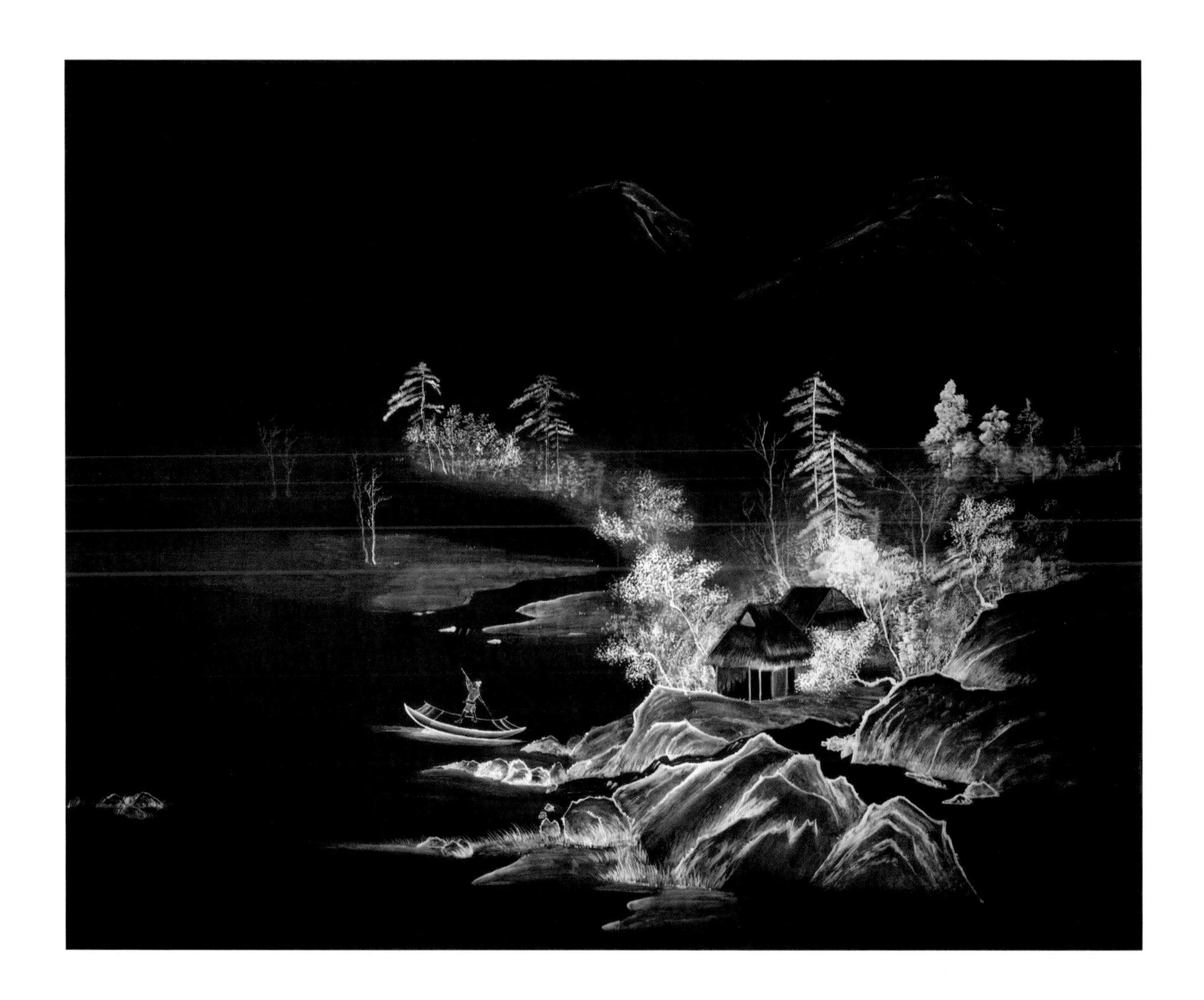

The living room uses silk wallpaper of gold foil paint. Silk is an ideal material for interior furnishing given its soft touch and the slightly shiny pearl white surface. The room's colour scheme involves black, gold and apricot, the combination of which feels gentle and elegant, and this, together with the carefully crated furniture, create a sense of modern luxury but in a accessible way.

客厅选用了金箔手绘的丝绸墙纸装饰背景，丝绸材质表层有轻微的珍珠光泽，质感较柔和适宜用于室内装饰。空间以黑、金、杏三色为主色调，色泽温润雅致，结合做工精细而有格调的家具饰品，共同营造了现代轻奢的氛围感。

The space is inspired by the shapes and mountain and louts. The guests will first be greeted by light blue coloured lotus lamp and oil painting of continuous mountains, beyond them the cerulean blue background colour-matching with the carpet of water-coloured louts, giving the space a unique world view described in Chinese landscape poetry.

空间营造上以山峦和莲花为主，映入眼帘的是淡蓝色莲花灯和山峦油画，而天青色的大色块背景又与水墨的荷花地毯遥相呼应；一深一浅，一天一地，浑然一体。

IMAGE © CZDF
图片来自春在东方品牌

IMAGE © CZDF
图片来自春在东方品牌

Before the Western Han Dynasty (206BC-9AD), the Eastern and Western cultures had collided and merged through the Silk Roads. After the Emperor Wu of Han Dynasty, the Silk Roads in the Sea deepened the integration of Eastern and Western cultures. In fact, the mutual absorption of Eastern and Western cultures far exceeded our imagination. East and West culture in the development of the entire era of the timeline is on the inclusiveness. Elegant Oriental style and modern Western fashion flourish in our living life. Especially in today's blending of traditional culture and foreign aesthetic, CZDF brand's "East Love West Rhyme" is a Neo-Chinese style way of life under the westward spread of Eastern culture.

西汉以前，东西方文化已通过丝绸之路碰撞和交融，汉武帝以后，海上丝绸之路更加深了东西方文化的融合，事实上，东西方文化的相互吸纳远远超过我们的想象。东西方文化在整个时代发展的时间轴上不断兼容并蓄、水乳交融、共舞争艳，儒雅的东方风格与摩登时尚的西方风格共同在我们的生活空间中活色生香。尤其是在今天的传统文化与西式审美的共融时代，“春在东方”品牌旗下的“东情西韵”是东学西渐下的新中式生活心法。

IMAGE © CZDF
图片来自春在东方品牌

IMAGE © CZDF
图片来自春在东方品牌

The ecological space with natural simplicity is achieved through the mottled wall and the Chinese style furniture.

斑驳而带有历史感的墙面、独具韵味的中式家具共同营造了自然古朴的原生态空间。

IMAGE © CZDF
图片来自春在东方品牌

"No Boundaries" focuses on users. It perceives senses and intends to create a large home furnishing lifestyle brand covering the color tone of the whole latitude.

东方［无界］以用户为中心，感知眼耳鼻舌身意，用心营造一个囊括色声香味触法的全纬度大家居生活方式品牌。

Ancient Chinese ink painting is circulated in the world, and indifferent tone rings through thousands of years. The Chinese literati with their delicate pen draw "Sparse shadows reflect horizontally in the clear and shallow water, and scented fragrance floats around under the dusk moon". Lotus and moonlight never had any noise. Put into few slice of tea, raise tea cup, slowly pour into the boiling water.

古韵墨画流传于世，淡然音律响彻万年，华夏文人，用他们纤弱细腻的笔，绘出"疏影横斜水清浅，暗香浮动月黄昏"。荷塘月色，青韵化蝶，不曾沾染过任何喧嚣，放入几片茶花，玉手提起茶盏，缓缓地倒入刚刚煮沸的春水。

IMAGE © CZDF
图片来自春在东方品牌

IMAGE © CZDF
图片来自春在东方品牌

5.2 HALLWAY DISPLAY
玄关赏析

Design

IMAGE © CZDF
图片来自春在东方品牌

5.3 PRODUCT DISPLAY

单品赏析

IMAGE © CZDF
图片来自春在东方品牌

01 XP17014 Leisure chair

Size l 60cm × 64.5cm × 76cm
Material l Lether / Zelkova

XP17014 休闲椅

尺寸 | 60cm×64.5cm×76cm
材料 | 真皮 / 榉木

02 Six pottery figures of musician

Size l 10cm × 8cm × 16cm
Material l Ceramic / Wooden base

六人坐乐俑

尺寸 | 10cm×8cm×16cm
材料 | 陶瓷 / 木底座

03 Table with pointed legs and nipple-pattered jade top

Size l 50cm × 50cm × 44cm
Material l Metal / Jade / Bronze legs

箭腿棕乳钉玉璧茶几

尺寸 | 50cm×50cm×44cm
材料 | 金属 / 玉石 / 铜脚

04 Rectangular box wiped in copper sheets, with stand

Size l 82cm × 62cm × 48cm
Material l Copper sheet / wood

长方形铜皮印箱带架子

尺寸 | 82cm×62cm×48cm
材料 | 铜皮 / 木质

05 XP17039 Closet

Size I 120cm×40cm×170cm
Material I Matte Jacobean wood stain /
Closet door upholstery

XP17039 高柜

尺寸 | 120cm×40cm×170cm
材料 | 黑橡哑光开放漆 / 柜门扪布

06 Over the boundary-PH980-5B

Size I 71cm×18cm×40cm
Material I Blank sand paint / Bronze / Matte white

墙外 -PH980-5B

尺寸 | 71cm×18cm×40cm
材料 | 砂纹黑 / 铜色 / 无光白

07 Paired lying Lion figures

Size I 50cm×50cm×150cm
Material I Ceramic / Pseudo-classical process

卧对狮

尺寸 | 8 寸
材料 | 陶瓷 / 仿古工艺

08 Round table made of copper sheets (L / S)

Size I 90cm×90cm×35cm
60cm×60cm×43cm
Material I Oriented Strand Board /
Copper sheet / iron-made legs

铜皮圆茶几（大 / 小）

尺寸 | 90cm×90cm×35cm（大）
60cm×60cm×43cm（小）
材料 | 奥松板 / 铜皮 / 艺腿

09 Story of Empty-city Stratagem painted in contract colours, Kintsugi blue-and-white porcelain plate, JZJS007 (wooden stand included)

Size I H:44cm D:44cm
Material I Porcelain / Kintsugi

空城计斗彩、青花双拼金缮盘 JZJS007(附实木盘架)

尺寸 | H:44cm D:44cm
材料 | 瓷器 / 金缮

10 Bamboo and Cicadidae Serious PH309-13A

Size I 61cm×15cm×65cm
Material I Blank / Bronze, Volakas base

竹蝉 – 嘻 PH309–13A

尺寸 | 61cm×15cm×65cm
材料 | 黑色 / 铜色，爵士白底座

11 Kintsugi blue-and-white porcelain gourd-shaped Vessel painted with green birds in bamboo forest JZJS009

Size I H:33cm D:23cm W:23cm
Material I Porcelain / Kintsugi

青花竹林翠鸟金缮东瓜坛 JZJS009

尺寸 | H:33cm D:23cm W:23cm
材料 | 瓷器 / 金缮

12 Taihu Scholar Rock-PH841-1B

Size I 47cm×17cm×47.5cm
Material I Brushed titanium / Emerald green / Marble base

太湖石记 –PH841–1B

尺寸 | 47cm×17cm×47.5cm
材料 | 钛金拉丝 / 翡翠绿 / 大理石座

VISUAL ART

视觉艺术

"Autumn wind is bleak, and life has come to the period when maple leaves begin to turn red. The rest of the seasons is only winter. However, life is just a cycle of four seasons. After winter, life is over. No matter how much storage, it is useless. When young, everyone should be hard. But no matter what the results are, people are old, and have the right to rest and to enjoy time with their grandchildren around the knee. Autumn beauty of life, should not be sad to remember, not with the unwillingness to continue to struggle, but to enjoy thoroughly."

Yutang Lin "The Art of Life

"秋风萧瑟，人生已到枫叶初红的时期。余下的季节只有冬。但人生只有一个四季的轮回。冬天过了，人生也就结束了。贮藏再多，也没有用武之地。每个人年轻的时候，是应该拼搏的，但无论结果怎样，人到老了，都有休息的权利，有儿女绕膝得享天伦之乐的时刻。人生秋天的美丽，不该带着伤感去缅怀，不该带着不甘继续奋斗，到了这个季节，就美丽地享受一次。"

林语堂《生活的艺术》

The Art of Life
生活的艺术

In his book "Art of Living," Mr. Lin tells us that, how to improve the quality of life apart from our work and responsibilities, to have a great taste of life, which is to enjoy life. People walking in a hurry and shuttling in the streets, are exhausted with life. In the eyes of many people, life gradually becomes a result or a goal. However, life should be enjoyed in peace and at ease.

Life is in thousands of postures. Will apperceptions of a sensitive heart be different from ordinary people? When thinking about how to make life not be a burden and not become the roller grinding our talent, but stimulate our creativity and beauty. Only to maximize the adaptation of nature can it achieve the spirit of freedom and liberation. Know the whole world from a small flower. In our daily life there are a lot of subtleties and unspeakable things, such as season, tide, impermanence of things, and so on. The unexpected experience is so fun from the nature. Life is art, and art also is life. Advocate the art of life, but also pursue artistic life. Since in ordinary life art is everywhere.

"Everyone is an artist" — German artist Joseph Beuys once said. Art is from life and higher than life. Art is from us both far and close, for that poetry and painting is art, tea is art, and in fact life is also an art. Life is mixed feeling and reflected in the daily trivia such as eating, walking, working, and studying. Comprehending art in life will make daily trivia no longer monotonous.

The purest beauty is from the nature. What we pursue in fact has been in the natural good fortune and quietly blooming. This stems from the attitude of life, so that art is as natural as breathing. The endless inspiration and encouragement make resource from trees that reflect the color of quiet blue sky, or a posture of blooming flowers in the spring which change and record our rich inspiration. We will consider a corner in the nature as a place for meditation and self-cultivation, to explore hidden taste between life and art in the balance.

林语堂先生在《生活的艺术》一书中告诉我们，工作之余、责任之外如何提高生活的质量、过上有品位的生活，即享受生活。人们行走匆匆，穿梭于街头巷尾，为了生活而疲劳奔命。在很多人眼里，生活，逐渐变成了一个结果，一个目标，很多人为了生活而生活。然而生活应当是快乐的，人们应当在生活中享受生活，和平地生活，安然处之，怡然自得。

生活千姿百态，一颗敏感的心所感是否与常人不同？当去思考怎么样让生活不成为一种负担，不变成磨碎了我们才华的碾子，并且能激发我们的创意跟美感。当乘物以游心，一花一世界。平常生活中的思绪流淌，很多微妙的、不可言说的东西，比如时令、潮汐、阴晴圆缺等来自自然的不期然凝聚成的点滴心得，妙趣横生。生活如艺术，艺术亦如生活，推崇生活化的艺术，也追求艺术化的生活，因为平凡生活里，艺术无处不在。

“人人都是艺术家”——德国艺术家约瑟夫·波伊斯（Joseph Beuys）曾说过这样一句话。艺术源于生活而高于生活，艺术离我们既远又很近，琴棋书画是艺术，品茶是艺术，其实生活同样也是一门艺术。生活是柴米油盐，五味杂陈，吃饭、步行、工作、学习，在生活中感悟艺术，才会使得这些日常琐事不再单调。

最纯粹的美，源于自然。我们追求的一切，梦中渴慕的、心中向往的，其实都已在自然的造化中，静静绽放美丽。这源于生命的姿态，让艺术像呼吸一样自然。给予我们源源不绝的激励与鼓舞的，也许是树木映着宁静蓝天的颜色，或是一朵花在春天绽放的姿态，幻化并记录着我们灵感最丰富的季节。我们则将这一隅自然视作一处静心之所，修行自我之所，去发掘那藏匿在生活和艺术的平衡之间的人生品位。

Record the artistic way of life
——记录艺术的生活

IMAGE © BOKING ART
图片来自铂晶艺术品牌

IMAGE © BOKING ART
图片来自铂晶艺术品牌

Visual Art Gesture
视觉艺术的姿态

From the beginning of human culture to today, human convey information through the visual image created by their own. Visual communication has always been the basic means of communication between people. These visual images are what we call "visual art" today. That is, with some materials, shaping visual art image of the plastic arts can be viewed by people. Modeling technique varied and the art image forms include sculpture, architectural art, decorative art and crafts and so on. Among them, the basic elements include: lines, shapes, light and dark, color, texture, space. These form the basis of a work. Design principles include: layout, contrast, rhythm, balance, unity. They are the principles and methods used by artists to organize and use basic elements to communicate meaning. Sculpture art is a kind of plastic arts, but also the manifestation of magic art form, also known as sculpture, sculpture and shaping in general. To create a visual and palatable artistic image with a certain space, in order to reflect the social life, to express the artist's aesthetic feeling, aesthetic emotion and aesthetic ideal, with the help of plastic (such as gypsum, resin, clay, etc.) or carving (such as metal, wood, stone, etc.).

There are thousands of objects in the world, which are diverse in content selection. Ranging from cosmic astrology to proton structure, from conscious subject reproduction to randomness composition, as long as there is a meaningful form, it can be expressed in the sculpture. Beautiful and mysterious crystal glass art work is very simple and pure in shape. If you want to create a unique color and lighting effects, you need to go through a series of complex processes. Cutting, laminating, carving and refining solid-state glass is critical for each step. Creating a variety of internal geometries patterns depends on how the array is arranged. It produces semi-permeable, frosted, faded halo and other more advanced texture after exposure to light, as if a processing of light to absorb the magic stone, and the light is locked in them. Appearance of the concise and complex internal contradictions constitutes a distinct body. The glass also has transparent and translucent properties, the perfect metaphor of the complexity and contradictions of life, making people read the aftertaste for a long time, with impulse to touch and play.

When growing-disappeared hand-made comes back to the contemporary society where everyone's eyes are focused on form and idea over all the things, the sculpture is not only a return to manual labor, but also deepest feelings of the most sincere arousal of human heart. These crystal glass works contain much principle of form beauty and enduringly form an irresistible "beauty" force. Its shape comes from accumulation of daily life, and the performance reflects a strong interest of life. It is filled with a particular era of secular life to embody personalized image. The way is rough yet concise, and fully demonstrates the beauty of nature and harmony.

从人类文化的开始到今天，人类通过自身创造的视觉形象来传达信息，视觉传达一直是人与人之间相互交流的基本手段。这些视觉形象在今天被我们称之为“视觉艺术”，即是用一定的物质材料，塑造可为人观看的直观艺术形象的造型艺术。其造型手法多种多样，所表现出来的艺术形象包括雕塑、建筑艺术、装饰艺术和工艺品等。其中，基本元素包括：线条、形状、明暗、色彩、质感、空间。它们是构成一件作品的基础。设计原则包括：布局、对比、节奏、平衡、统一。它们是艺术家用来组织和运用基本元素传达意义的原则和方法。雕塑艺术，是造型艺术的一种，但同时也是魔幻艺术的表现形式，又称雕刻，是雕刻和塑造的总称。以可塑的（如石膏、树脂、黏土等）或可雕刻的（如金属、木材、石头等）材料，创造出具有一定空间的可视、可触的艺术形象，借以反映社会生活，表达艺术家的审美感受、审美情感、审美理想的艺术。

世间物象千千万万、数不胜数，这一切都为创作内容选择带来多样化，艺术家经过筛选、处理和加工，大至宇宙星象，小到质子结构，从有意识的主题再现，到随意性的形态构成，只要具备有意味的形式，均可浓缩于雕塑作品之中。美丽而神奇的水晶玻璃艺术作品，虽然外形都很简单、纯净，但想要创造出内部特有的色彩及光照效果，需要经过一系列复杂的工艺才能实现。对固态玻璃的切割、层压、雕琢与提炼，每一步骤都很关键，根据排列方式的不同从而创造出各种内部几何图案。用光线照射后更是产生半透、磨砂、褪晕等更高级的质感，仿佛一个对光线进行加工吸收的魔法石，将光线锁在其中。外表的简洁与内部的复杂构成鲜明的矛盾体，玻璃同样具有透明与半透明两种属性，完美地隐喻了生命体的复杂性与矛盾性，让人看过后回味良久，甚至很有抚摸把玩的冲动。

当日渐消失的手工制作在当代多元文化对形式和观念的玩味高于一切的时候，雕塑手作不仅仅是一种对手工劳作的回归，还有对人类内心深处情感的最恳切的唤起。这些水晶玻璃作品本身蕴含着丰富的形式美法则，经久不衰，形成了一种不可抗拒的“美”的力量。其造型来源于日常生活积累，表现的是浓厚生活情趣，充盈着特定时代世俗生活气息上体现着个性化的形象，手法粗犷大气中又不失简练，充分表现出了自然与和谐之美。

6

BOKING ART OF GREAT PURITY

铂晶艺术

IMAGE © BOKING ART
图片来自铂晶艺术品牌

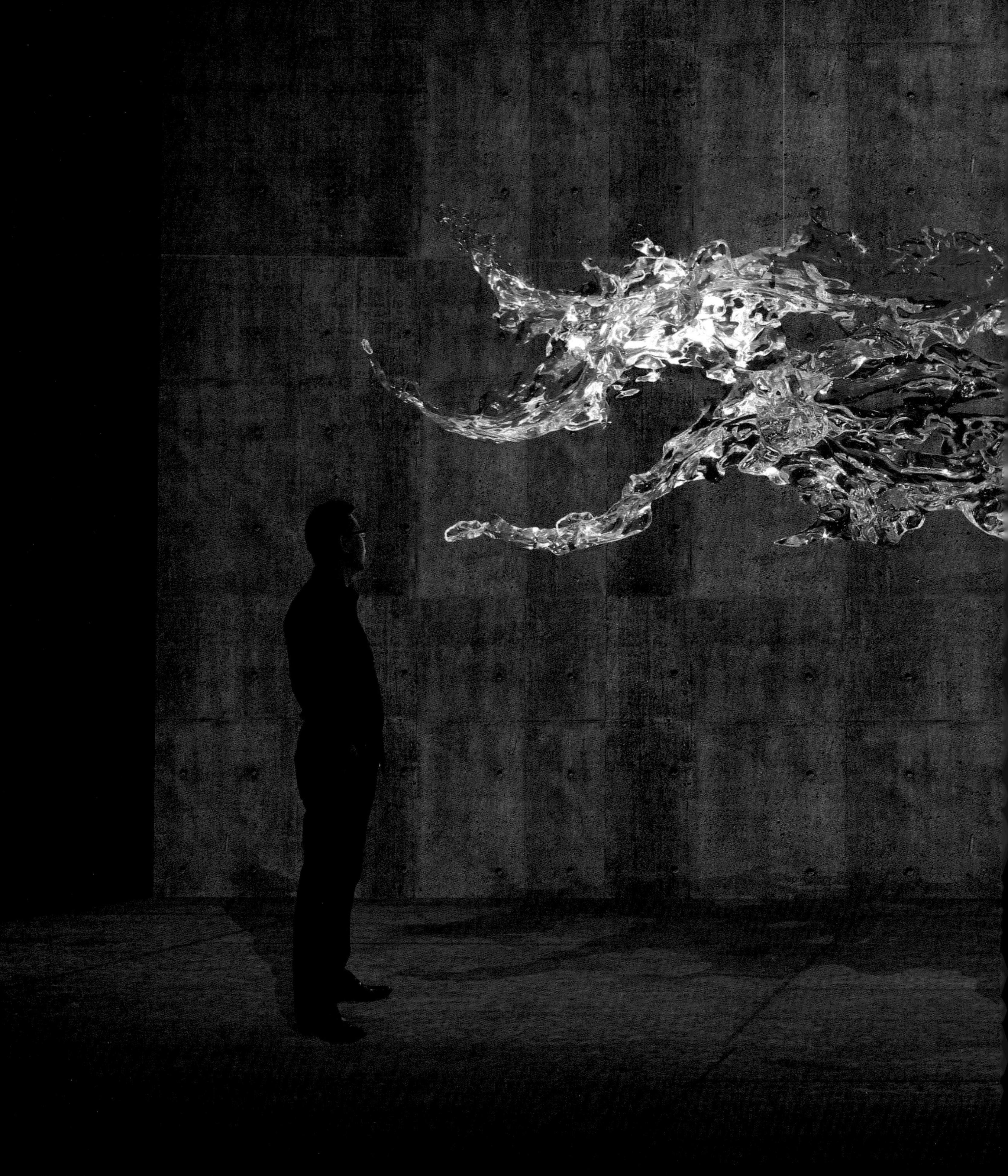

The highest excellence in the like that of water.

"The best of man is like water, which benefits all things, but strives for nothing."
In the Tao Te Ching of Laozi.

4500mm × 1300mm × 1000mm

As Good As Water

[上善若水]

最高境界的善行就像水的品性一样，泽被万物而不争名利。

语出《老子》：“上善若水，水善利万物而不争。”

IMAGE © BOKING ART
图片来自铂晶艺术品牌

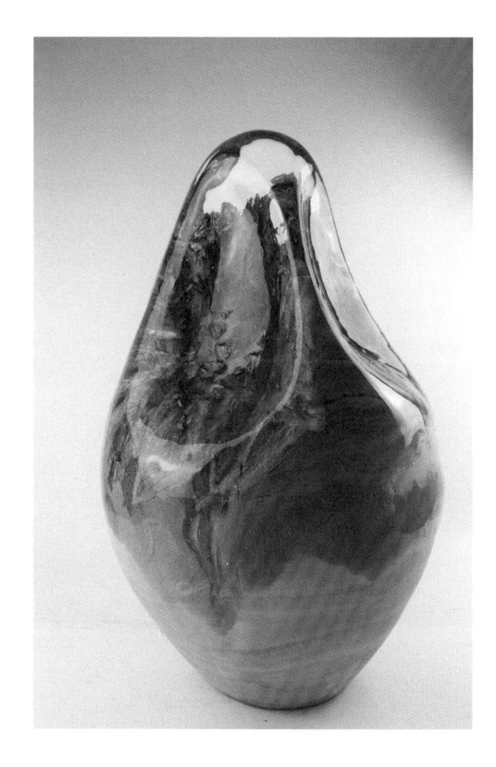

When BOKING ART met craftsmen, You will see it blooming out of another style.

当铂晶艺术遇到了手工艺人，
你会看见它绽放出另一种风采。

Flexible degree, simplified appropriately, bending and straightening suitably, combining force with mercy, which is also full of music rhythm.

张弛有度，繁简得宜，曲直相宜，刚柔相济，这其中也充满着音乐的节律感。

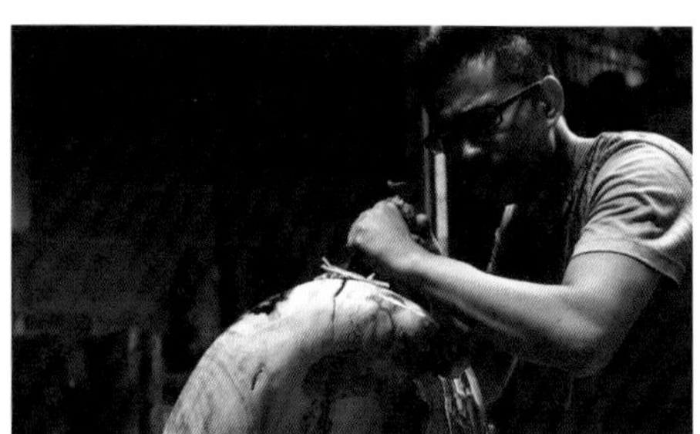

Laozi said: "Tao that can be described is not universal and eternal Tao. Name that can be named is not universal and eternal Name." We can understand that the Tao is natural and unspeakable. Laozi advocated the harmony of nature with "Great music has the faintest notes, great form is beyond shape", and was extended by the artists of the BOKING ART as advocating naturalism without leaving too many artificial carvings. It is just the beauty to the point.

老子曾言："道可道，非常道，名可名，非常名。"我们可以理解为，道是自然的，是不可言说的。老子提倡"大音希声""大象无形"的自然和谐境界，被铂晶艺术家们进行引申，成为他们崇尚自然天成，而不留过多人工雕琢痕迹的一种恰到好处的美的境界。

图片来自铂晶艺术品牌

Three-Dimensional Ink & Wash

Breeze, Moon, Cloud and Zen

195mm×195mm×510mm

195mm×195mm×570mm

[三维水墨]

清风·明月·云水·禅心

IMAGE © BOKING ART
图片来自铂晶艺术品牌

300mm×300mm×400mm

将原木／枯枝用科技的手法永久封存在铂晶里，再经过细细的手工打磨成可以坐的墩子，意外与惊喜就在这里。

The log and deadwood with technology approach is sealed permanently in the platinum crystal, then after careful hand-polishing, it turns into unique stool. Surprises are here.

IMAGE © BOKING ART
图片来自铂晶艺术品牌

510mm×470mm×660mm

The soothing water flow gives a sense of inward condensation. Between ideological and practical the visual center is exceptionally vivid.

流转舒缓，给人一种向内凝聚的感觉。虚实之间，使得视觉中心分外传神。

360mm×360mm×600mm

Φ1100mm

IMAGE © BOKING ART
图片来自铂晶艺术品牌

IMAGE © BOKING ART
图片来自铂晶艺术品牌

Low-back Chair

低背椅

Wood · Luxury

Using high-tech transparent material moulding process artists present the "new luxury" doctrine. Each work is unique, made of rotten wood, dead branches, leaves that after years of baptism already left with time traces. Artists use artisan spirit to create a series of works of art, so dead wood, dead branches and leaves breathe, glowing new lives. Time seems to be still, but life has been sublimated.

木·华

采用高科技透明材料成型工艺与大自然、艺术家共同碰撞出的“新奢华”主义。

每一件作品都是独一无二的，经过岁月洗礼的朽木、枯枝、树叶，早已留下时间记忆的痕迹，艺术家运用匠人精神创作出的系列艺术品，使朽木、枯枝、树叶有了呼吸，焕发出新的生命，时间仿佛静止，而人生却得到了升华。

IMAGE © BOKING ART

图片来自铂晶艺术品牌

3200mm×500mm×800mm

铂晶太湖石

Boking Taihu Stone

600mm×550mm×1700mm

IMAGE © BOKING ART
图片来自铂晶艺术品牌

She looked around, static yet dynamic. It is so lifelike that we almost hear breathing.

Carving in unique perspective rules,

pay attention to metaphysical freehand spirit. Without too formal decoration.

She only has the real expression, lyrical body language, conveying to us indifferent life posture.

她左顾右盼，静中含动，

栩栩如生，几乎令人听见呼吸。

雕刻独特的透视法则，

注重形而上的写意精神，

没有过于形式化的装饰。

她有的只是真实的表情，

抒情的肢体语言，

传达给我们淡然的人生姿态。

270mm×320mm×980mm

500mm×550mm×1700mm

IMAGE © BOKING ART
图片来自铂晶艺术品牌

520mm×220mm×820mm

IMAGE © BOKING ART
图片来自铂晶艺术品牌

Vast Landscape

Dialogue between contemporary and ancient people across time and space, pursuit of Tao & Qi in Chinese classical landscape through artistic language.

千里江山

当代与古人穿越时空的对话，以当代的艺术语言追求中国古典山水的“道”与“气”。

800mm×600mm×5000mm

CLASSICAL ARISTOCRACY

古典贵族

Quality Life
品质生活

In 2010, "Downton Abbey" was broadcasted in ITV. Subsequently, the drama about the life of the British noble, appeared in the Emmy and Golden Globe Awards and other list of winners. The drama's success depends on, in addition to exciting story, vivid Aristocratic life, overflowing aristocratic spirit and classical elegance of the interior decoration which is the absolute highlight.

Pay attention to the aristocratic style, in particular value the style and quality of life. Everyone may look forward to the elegant noble life, wearing high-level customization of the elegant dress, elegantly bringing up a glass of red wine or champagne to attend a variety of social Cocktail party, or dressed in extreme-perfection handsome riding habit in leisure hunting time or enjoying the afternoon tea with British rituals in the elegantly furnished residence. The greatest charm of noble life is not extravagance, nor a big house or the number of servants, but it is the spirit revealed from the bones, the details of life, that indeliberately and naturally showing the noble aristocratic temperament.

In "Downton Abbey" the scene that manservant irons newspaper for earl impresses audience so deeply for the exquisite life. Such small thing is so delicate, let alone its daily residential decoration.

2010年，《唐顿庄园》登录ITV。随后，这部讲述英国贵族生活的电视剧，出现在包括艾美奖和金球奖在内的各项获奖名单中，这部剧集的成功，除了精彩的剧情之外，其跃于屏上的贵族式生活、满溢的贵族精神和经典优雅的室内装饰也是绝对的亮点。

讲究格调的贵族，尤其看重生活方式和生活质量，每个人心中或许都憧憬过贵族那般考究的生活，或穿着高级定制的精美礼服，优雅地端起一杯红酒或香槟，出席各种名流的社交酒会，或身着极致完美的帅气骑装以狩猎消遣闲暇时光，或在装潢考究的宅邸内优雅地享受那充斥着仪式感的英式下午茶。熟知贵族生活的人都知道，其最大的魅力并不在于标榜奢华贵气，也不在于是否有大房子或者多少仆人鞍前马后，而在于那从骨子里透出来的一种精神，一种生活细节和教养，不必刻意，却自然流露出的翩翩贵族气质。

《唐顿庄园》剧中男仆为伯爵一家熨烫报纸的细节让观众记忆深刻，并不由得感叹其生活的精致与讲究，小到看报的情景尚且如此细腻，更何况是其日常起居的住宅装饰呢？

Fashion is a trend, and aristocracy is a kind of accumulation
时尚是一种潮流，贵族是一种积淀

IMAGE © DI GAO MEI JU
图片来自帝高美居品牌

IMAGE © DI GAO MEI JU
图片来自蒂高美居品牌

European and American classical style

欧美古典风情

European classical decoration style

European classical charm lies in its unique historical traces which reflect the elegant timeless bearing on behalf of an excellent quality of life of owner.

European classical style is evolved from the aristocratic lifestyle, which contains elements just to meet the demand for lifestyle of current cultural bourgeois, namely: a sense of culture, a sense of nobility, no lack of a sense of freedom and mood, but also the pursuit of history and culture, which is not only reflected in the ornaments on the antique works of art, but also reflected in the decoration of various antique tiles, stone preferences and the pursuit of a variety of antique processes. In general, the classical aristocratic style of decoration is elegant and full of history.

European classical style with gorgeous decoration, strong colors, beautiful shape to achieve the elegant decoration effect, vividly embodies the rich cultural heritage of European culture. The style presentation is extremely particular, giving the impression of dignified and elegant, noble and gorgeous, with a strong cultural atmosphere. Furniture matching generally uses large fine furniture, together with the exquisite carving. The overall effects create a gorgeous, noble, warm feeling. It often uses golden and brown accessories to bring out the noble and elegant classic of furniture. In color, it often uses white or yellow series as the basis, with dark green, dark brown, gold, etc., showing the classical European style luxury temperament. In material, it generally uses cherry wood, walnut and other high-end solid wood, showing noble and elegant aristocratic temperament.

欧式古典装饰风格

欧式古典的魅力，在于其独具历史岁月的痕迹，其体现出的优雅隽永的气度代表主人的一种卓越的生活品位。

欧式古典风格是贵族生活方式演变至今的一种形式，其所蕴含的元素也正好迎合了时下文化资产者对生活方式的需求，即：有文化感、有贵气感，不能缺乏自在感与情调感的同时，也追求历史文化感，这不仅反映在装饰品上对仿古艺术品的喜爱，同时也反映在装修上对各种仿古墙地砖、石材的偏爱和对各种仿旧工艺的追求上。总体而言，古典贵族风格的装饰是典雅而富有历史气息的。

欧式古典风格以华丽的装饰、浓烈的色彩、精美的造型达到雍容华贵的装饰效果，淋漓尽致地体现了欧洲文化丰富的艺术底蕴。在造型上极其讲究，给人的感觉端庄典雅、高贵华丽，具有浓厚的文化气息。在家具选配上，一般采用宽大精美的家具，配以精致的雕刻，整体营造出一种华丽、高贵、温馨的感觉。在配饰上，常采用金黄色和棕色的配饰衬托出古典家具的高贵与优雅。在色彩上，经常以白色系或黄色系为基础，搭配墨绿色、深棕色、金色等，表现出古典欧式风格的华贵气质。在材质上，一般采用樱桃木、胡桃木等高档实木，表现出高贵典雅的贵族气质。

American classical decoration style

American classical style is rooted in European culture, and it abandoned the Baroque and Rococo style which pursued novel and flashy, and was based on a new understanding of classic, emphasizing concision, clear lines and elegant, decent decoration.

American classical style formed a mixed one in the transmission of European culture and combined with the characteristics of their own culture of the United States.

Referring to elements and characteristics of classical and neo-classical style is a major bright spot of American classical style. Decoration material of American classical style is mainly hard and gorgeous. Bright color is less. Curtains and wallpaper also choose soft colors. To create a warm classical temperament, it abandons the magnificent European-style temperament to focus on practicality and coordination of the overall home furnishing.

American classical style decoration considers the various uses of each space. The American classical style living room is relatively low and gentle which selects comfortable and soft material and pays attention to family atmosphere. American classical style furnishings are also very distinctive. The main colors are black, dark red, brown and other dark colors. American classical style furniture prefers darker color that looks stable and elegant. The bed has a high column and veiling. The graceful bed mantle can present elegant beauty of the American classical style. Also, American classical style chair highlights the characteristics of the "Queen Anna", and American furniture carving through aging treatment, wormhole, erosion and other embellishments displays the beauty of American classic.

美式古典装饰风格

美式古典风格植根于欧洲文化，它摒弃了巴洛克和洛可可风格所追求的新奇和浮华，建立在一种对古典的新的认识基础上，强调简洁、明晰的线条和优雅、得体有度的装饰。

美式古典风格是在传承欧式文化的基础上结合了美国自身文化的特点，而形成的一种混合风格。

借鉴了古典主义和新古典主义风格的元素和特点是美式古典风格的一大亮点。美式古典风格装修材质主要为坚硬、华丽的材质，亮色较少，窗帘和壁纸也多选用柔和色彩，营造温馨的古典气质。摒弃了欧式的富丽堂皇的气质，转而更注重家居整体的实用和协调。

美式古典风格装修更考虑每个空间的多样用处。美式古典风格起居室较低矮平和，起居室选材也多采用舒适柔软的材质，更注重家庭氛围。美式古典风格装修中的家具也很有特色，主色调是黑、暗红、褐色及深色的美式古典风格家具，在家中更显稳重优雅。其卧床有高柱和帐幔，曼妙的床幔可以看出美式古典风格的优雅美。美式古典风格的椅子凸显了“安娜女王”的特点，而美式家具的雕刻在做旧、虫眼、侵蚀等的点缀下，尽显美式古典之美。

IMAGE © DI GAO MEI JU
图片来自蒂高美居品牌

Classic ornaments soft decoration matching skills

New classical furniture in recent years developed fast in domestic market showing a good momentum of development. Classical, neo-classical furniture of historical heritage, compared with modern plate-typed furniture, has a more profound cultural background. Therefore, the layout of classical home furnishings, including matching of furniture, lighting and ornaments, needs more ingenuity and more skills.

古典饰品软装搭配技巧

新古典家具近几年如雨后春笋，在内销市场迅速蹿红，呈现出良好的发展势头。古典、新古典家具因为历史传承的原因，相较现代板式家具，拥有更为深厚的文化底蕴。因此，古典家装的布置，包括家具搭配、灯光以及饰品的摆设，都更需匠心，更需功底。

Choice of ornaments color

Furniture accessories are embellishment for the furniture, making furniture more luxurious and high-end, and to express foil effect of ornaments is the ability to control the performance of the entire space for designers. In the choice of ornaments the first is the same color matching, with the color to be similar or close, and furniture in harmony with the overall unity will produce the overall beauty.

Select ornaments similar to furniture

European, American classical furniture have their own unique technology and production processes. For each manufacturer the details of furniture such as hardware inlay and exquisite carving will be different, which inevitably requires furniture accessories in the style and appearance adapted to furniture. Similar carving patterns have a harmonious beauty. When European-style patterns match American-style furniture there will be the feeling of mistaken identity.

Distinguish the main color, sub-color and embellishment

Place ornaments in a bright layering. In an exhibition hall there should not be just one tone which is inevitably monotonous. There must be a sub-tone in addition to the main color, thus to form a sense of hierarchy. With the main, sub-tone color, ornament exhibition hall is not perfect but also needs different types or colors of ornaments which have played a focal point in the entire exhibition. It is just like that dropping a stone in calm water and ripples make water surface dynamic.

Allocation in line with the aesthetic

The entire home furnishings should be like a set of traditional Chinese painting. It is a sort of skill from composition to coloring. The rationality placement of furniture affects the level; color harmony affects the appearance. The perfect matching of furniture and accessories is embodiment of texture and nobility. Accessories are usually placed in two ways: from point to line and then to the surface, or vice versa, from the surface to the line and then to the point. Designers can choose the way to display.

Beauty lies in the details

Ornaments are placed in the general practicing of the "Round and square with the pursuit of harmonious development" theory. Take bedside cabinet as an example. After choosing the concordant style, color and ornaments, if surface is a square, the shape of furniture should be round or oval, vice versa. The match between ornaments is also important. Pay attention to the height of the accessories: if furniture is upright and foursquare, the accessories should be flexible, or follow low-high-low or high-middle-low. If furniture is round, ornaments should be structured. Height difference should not be too large.

饰品颜色的选择

家具饰品是为家具做点缀，让家具更显华贵和品位，家具饰品的衬托作用如何表现得淋漓尽致，是设计者驾驭整个摆场能力的体现。在选择饰品时，首要的是同一色系的搭配，色系要相近或者相似，才能和家具在整体搭配上和谐统一，才会产生整体的美。

选择器型和家具相似的饰品

欧式、美式古典家具都有自己独特的工艺和生产程序，每个生产厂家在家具的细节如五金的镶嵌以及雕刻精细方面都会各有不同，这必然要求家具饰品在款式和外观器型上要和家具相适应。类似的雕刻图案会有和谐的美感，而欧式风格的图案与美式风格家具相配，则会有张冠李戴的感觉。

分清主色、次色及点缀

饰品的摆放要有明快的层次感，在一个展厅中，不能就一个色调，这样难免失之单调，一定要在主色调之外，安排一种次色调，从而形成层次感。有了主、次色调饰品的展厅还不算完美，还要有几件器型或者颜色不同于主次饰品的摆件，对整个展场起到“点睛”的作用。它们就像在平静的水面上扔下了一块石头，泛起的涟漪让水面富有动感。

合乎审美的配放

整个家装的饰品配放应该如同一幅国画，从构图到着色都是一种技巧。家具摆放的合理性影响层次；色彩的和谐性影响外观。家具和饰品完美的搭配是家具的质感和高贵的体现。通常饰品配放会按两种方式进行：或从点到线再到面，或反之，从面到线再到点。采用何种方式去做，要看设计者的思路。

美在于细节

在饰品摆放中，一般采用天圆地方或者天方地圆的方法。以床头柜为例，选好款式及色系与床头柜一致的饰品后，当床头柜的面是方形时，家具饰品的外形应该选圆形或者椭圆形；当床头柜的面是圆形时，则应该选方形的家具饰品。饰品之间的搭配同样重要，要注意从饰品的高度上分摆：家具方正，则饰品要灵活，可以“低高低”，也可以“高中低”，家具圆润，饰品则规整，高低差别不宜过大。

7

DI GAO MEI JU

蒂高美居

IMAGE © DI GAO MEI JU
图片来自蒂高美居品牌

IMAGE © DI GAO MEI JU
图片来自蒂高美居品牌

CREATIVE
walls
THE SCARPETTA FACTOR
Patricia Cornwell

Every detail of exquisite hand-painted patterns is superb. Exquisite lacquering technology and high color saturation make the fresh green striking and pure, as if a heartbeat of love.

精致的手工彩绘图案，一花一叶每一笔都勾勒得出神入化。精湛的漆工艺，极高的色彩饱和度，那抹清新的翠绿醒目又纯粹，好似一份怦然心动的爱情。

7.1 LIVING ROOM

客厅赏析

IMAGE © DI GAO MEI JU
图片来自蒂高美居品牌

DC Outlets

IMAGE © DI GAO MEI JU
图片来自蒂高美居品牌

THE EARTH

IMAGE © DI GAO MEI JU
图片来自蒂高美居品牌

Love of copper objects can be described as a complex. Childhood in a sense should be called the copper years. No toys, but pick up coins, copper and others from beach. Hide them in the pocket in addition to a few pieces of copper shells. Love its high-quality texture and heavy sense of reality when you place it in the hands, with a proud attitude such as gold to highlight bright light, but with a humble low-key attitude to interpret the dignity and ancient rhyme.

喜欢铜质物件，可谓一种情结。童年，从某种意义上来说，应该称之为铜年。没有玩具，就从河滩中拾的铜钱、铜板之类，再加上几枚铜弹壳，揣在兜里当宝贝显摆。爱它表里如一的质感和掂在手里沉甸甸的实感，带着一种高傲如金的姿态彰显耀眼光芒，又用一种谦卑的低调诠释出厚重和古韵。

7.2 DINING ROOM
餐厅赏析

IMAGE © DI GAO MEI JU
图片来自蒂高美居品牌

IMAGE © DI GAO MEI JU
图片来自蒂高美居品牌

Craft production of the restaurant furniture is meticulous, in which there are hand-painted applications, gold and silver platinum features selections. All is beautiful. As long as you spend some extra patience and use those seemingly unimportant props with the suitable shape and color, you can easily create a romantic sweet lovers' world!

本案餐厅的家具工艺制作处理得细致入微，手绘的涂装，金铂、银铂的特征选择皆美不胜收。只要花一些巧心思，利用那些看起来不重要的小道具，选对造型和颜色，就能轻易营造出一个浪漫甜蜜的二人世界！

7.3 PRODUCT DISPLAY

单品赏析

IMAGE © DI GAO MEI JU
图片来自蒂高美居品牌

It uses a large number of complex carving processes, partly pasting gold foil. Some combinations are also with a small amount of black trim, and even more elegant. It symbolizes the international low-key luxury, filled with taste and fashion. Through the space furnishing display, it awakens people to deepen feelings to the art of home life, and enjoy the extraordinary luxury and royal life.

采用大量复杂的雕刻工艺，局部贴金箔。有些组合还带有少量黑色镶边，雍容华贵。象征着国际化的低调奢华，弥漫着品位与潮流，透过空间陈设，唤醒人们深藏对居家生活的艺术情感，享受凝聚皇族超凡的奢华生活。

Ceramic inlay copper

Ceramic with copper is precision casting process which can create thin-walled and beautiful-outlined copper sculpture.

陶瓷镶铜

陶瓷配铜，属于精密铸造工艺，可以制造薄壁，勾勒出线条优美的铜饰雕塑。

Derived from the exotic Spanish style, it uses fine grinding and polishing process to forge a smooth and sharp metal texture.

源自于西班牙的异域风情，以精细的打磨和抛光工艺锻造出流畅锐利的金属质感。

Design is sourced from abroad collection of antique masterpieces and copy hand-painted pattern. Organic combination of two different elements with a thick and sharp copper texture and simple natural ceramic overflow thick classical culture.

设计源自国外珍藏的古董名著，临摹纯手工彩绘图案，敦厚锐利的铜质感与淳朴自然的陶瓷，两种不同元素有机结合，满溢浓浓的古典文化气息。

IMAGE © DI GAO MEI JU
图片来自蒂高美居品牌

02 Size | 24.5cm × 18cm × 58.5cm
尺寸 | 24.5cm×18cm×58.5cm

01 Size | 30.5cm × 22.5cm × 35.5cm
尺寸 | 30.5cm×22.5cm×35.5cm

03 Size | 34cm × 34cm × 72.5cm
尺寸 | 34cm×34cm×72.5cm

04 Size | 50cm × 20cm × 68.5cm
尺寸 | 50cm×20cm×68.5cm

05 Size | 26cm × 26cm × 33.5cm
尺寸 | 26cm×26cm×33.5cm

IMAGE © DI GAO MEI JU
图片来自蒂高美居品牌

06 Blue and white porcelain candlestick decoration

Although the color of blue and white porcelain is single, but it does not make people feel monotonous. Skilled porcelain painting artists select whether concentrated light or dense wiping to express smartness or deliberately seek structure or use little space to complete. All could be called "write like an angel" and is hard to express its beauty completely.

青花瓷烛台摆件

青花瓷虽然色彩单一，却并不让人感觉单调。功力深厚的瓷绘艺人，或浓施淡抹，笔触灵动；或刻意求工，层次分明；或寥寥数笔，一气呵成；线条曲直，蜿蜒飘逸，几分精细，几分狂放；堪称妙笔生花，美不可言。

07 Size | 25cm×25cm×63cm

尺寸 | 25cm×25cm×63cm

08 Size | 23.5cm×21cm×45cm

尺寸 | 23.5cm×21cm×45cm

IMAGE © DI GAO MEI JU

图片来自蒂高美居品牌

09 Copper with ceramic porch

Copper sculpture with complex shape, fine pattern and multifarious process, highlights the extraordinary quality of the European palace furniture. In the entrance hall it manufactures magnificent concrete microcosm, giving a stunning sense.

铜配陶瓷玄关

形状复杂、花纹精细、工序繁杂的铜饰雕塑，彰显欧式宫廷家具饰品的非凡品质，于玄关处制造堂皇空间里具体而微的缩影，给人一进门的惊艳感。

10 Deer head wall clock

This is a "wall clock" of Rococo style, with a beautiful deer head and asymmetrical leaf decoration around the dial.

鹿头墙钟

这是一款洛可可风格的“墙钟”，带有优美的鹿头，且绕表盘缀有不对称的叶形装饰。

11 "Gigant" clock

"Gigant" clock is a very beautiful Louis XVI style table clock. The model is casted in the shape of building, and there are a lot of branch decoration on the front and side.

“贾甘特”台钟

“贾甘特”台钟是非常漂亮的路易十六风格的台钟，模型以建筑形状铸，正面和侧面有很多花枝雕饰。

Retro and classic merged into the superb skills

复古与经典融于精湛技艺

01 Hand-painting

Hand-painting process is complex. Brushing several times, elution and a specific halo technology in the production process can make the color of home accessories better integrated, and increase the rich layering for the pattern. Also, its use is very flexible, matching the temperament of the corresponding pattern according to the different styles of furniture, but also can choose brushwork, watercolor, oil painting and other options in the style of painting. The series of home accessories are decorated with exquisite hand-painting for the surface of the plain wood and ceramic home accessories to put on elegant clothes and add a rich sense of emotional elements and aesthetic appeal.

手工彩绘

手工彩绘工艺繁复，其制作过程中多次的涂刷、洗脱及特定的晕染技术，能够使家居饰品的颜色更好地融合，并为图案增加丰富的层次感。其运用非常灵活，可根据不同风格的家具配搭与其气质相应的图案，而在绘画风格上也有工笔画、水彩画、油画等多种选择。这一系列的家居饰品表面皆采用精致的手工彩绘，为表面平实的木质和陶瓷家居饰品穿上精美华衣，添加丰富的感性元素及审美情趣。

02 Ceramics

China is one of the earliest countries to apply pottery in the world, and Chinese porcelain has been highly respected by the world because of its high practicality and artistic quality. The series of ceramic products are concise and generous without losing elegance. Its shape, glaze and decoration have created an implication of art image.

陶瓷

中国是世界上最早应用陶器的国家之一，而中国瓷器因其极高的实用性和艺术性而备受世人的推崇。这一系列陶瓷制品，简约大方又不失典雅高贵，无论是其造型、釉色、装饰，皆创造了一个艺术形象的审美意蕴。

03 Pourie inlay technology

Famous master of the palace furniture of France, Pourie develops creatively exquisite inlay wood craft in which the metal plate and the tortoise shells overlap to cut into patterns and are embedded in the furniture surface, forming a Pourie aristocratic decoration style and becoming main style of the French Rococo period. Unfortunately, the process has defect that "patch may fall off with the passing of time".

The series of home accessories inherit precious craftsmanship, while improving it to a higher level. These products use hand-painting simulation process to restore the precious wood chip mosaic collage effect, and the original metal shell mosaic is updated to gold foil mosaic. It not only retains the beauty of the mosaic process, but also is easy for collection and maintenance and enhances the overall furniture value.

鲍里镶嵌技术

法国著名的宫廷家具大师鲍里创造性地发展精致的镶嵌细木工艺，将金属片与龟甲重叠在一起切成图案，镶嵌在家具表面，形成了一种鲍里式贵族化的装饰风格，并成为法国洛可可时期的主要风格。遗憾的是，这种工艺存在"贴片可能随着时间流逝而脱落"的缺陷。

而这一系列的家居饰品在传承珍贵工艺的同时，对其又进行了改良提升。这些产品中采用手工彩绘仿真工艺，还原名贵木材薄片的镶嵌拼贴效果，并将原来的金属龟甲镶嵌术升级为金箔镶嵌术，不仅保留镶嵌工艺的美感，易于收藏保养，更提升了家具的整体价值。

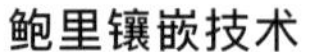

IMAGE © DI GAO MEI JU

图片来自蒂高美居品牌

04 Antique technology

Painting antique aging treatment is an important process of antique furniture, which uses a series of techniques for the new furniture to increase the ancient sense and to give a long history of imprinting. When restoring historical style it also adds unique human value to furniture. As "classic from the museum", "DI GAO MEI JU" home accessories use extensively painting antique craft to restore pieces of classical paintings and classic ornaments in museums in real arena, and this process is also perfectly shown in single item.

仿古技术

彩绘仿古做旧是仿古家具中的一项重要工艺，它运用一系列工艺手法为新家具增加古旧感，赋予家具漫长的历史印记，在还原历史风貌的同时，为家具附加独特的人文价值。作为"博物馆里走出的经典"，"蒂高美居"家居饰品中大量使用彩绘仿古工艺，还原一件件古典名画及博物馆里的经典饰品于现实舞台，而这一工艺也在此单品中得到了完美展现。

05 Blue and white porcelain

Blue and white porcelain as a representative of the history of Chinese ceramics, is a symbol of ancient Chinese cultural heritage. It acculturates profound Chinese culture into the West, and has an indelible role in the process of cultural exchanges between China and foreign countries. Blue and white colors in opposite, simple and bright, hearty and generous, is the beauty of elegant fresh. Blue and white record the classic tastes, and delicate lines with elegant structure construct classical artistic concept.

青花瓷

青花瓷作为中国陶瓷器史上的代表，是古老的中国文化底蕴的象征。它把博大精深的中国文化传入西方，在中外文化交流过程中有着不可磨灭的作用。蓝白两色对映，朴素明快，爽朗大方，具有典雅清新之美。涩白青花，记录经典的味道，细腻的线条，优雅的结构凝聚属于经典的气韵。

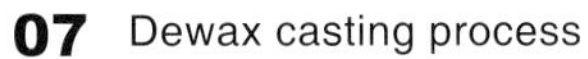

06 Gilding techniques

Inlay gold (or silver) is an ancient traditional craft. European palace home accessories are mostly gold foil surface, once exclusive for royal family and the nobility, a symbol of status and wealth. "DI GAO MEI JU" has inherited and improved gilding techniques on the basis of traditional process. The surface sprayed primer paint of the furniture has been sprayed a special background color in hand brush or spray, with a viscous water aqua in the background on the special treatment, and attached to the gold foil, creating a variety of artistic effects in close to the furniture accessories of the gold foil surface. Gilding technology allows the furniture accessories to maintain the luxury of the European-style palace expression. Gold foil and the surface of the furniture accessories are closely connected, illustrating the exquisiteness and luxury.

贴金（银）术

贴金（银）术是一项古老的传统工艺，欧式宫廷家居饰品表面大多都是金箔镶嵌，一度只是宫廷皇族及权贵们的专享之物，是地位与财富的象征。"蒂高美居"在传统工艺的基础上对贴金术进行了继承和改良。在已喷封底漆的家具饰品表面再手刷或喷涂一层特别的底色，用具有黏性的水剂在底色上进行特别处理，然后把金箔贴附其上，再在紧贴家具饰品表面的金箔表面做出各种艺术效果。贴金技术让这一家具饰品保持了欧式宫廷的奢华表情，金箔与家具饰品表面紧密连接，尽显精美与华贵。

07 Dewax casting process

To highlight the extraordinary quality of European-style palace furniture and to restore the real material of palace furniture, the luxury series of ceramic ornaments use mostly copper castings decorative style. Copper decorations often use dewax casting process. Cere method is a precision casting process which can create copper sculpture with thin-walled, complex shape, fine pattern, and complex process.

In the making process, casting process is used to make the paraffin wax model. The sculptured shell is made of quartz sand and other casting material. Then, the shell is dewaxed and baked again, and then the copper water is poured. Broken shell after copper water is cooling to complete copper products. Finally, the surface of copper is to deal with coloring, corrosion and other treatment, so as to complete the sculpture production. It decorates ceramic surface with exquisite bronze, highlighting the extraordinary identity and style of furniture immediately.

脱蜡铸铜

为彰显欧式宫廷家具饰品的非凡品质，还原宫廷家具的真材实料，这一奢华系列陶瓷饰品中大部分沿用了铜铸件的装饰风格。铜质装饰多采用脱蜡铸铜工艺，蜡模法属于精密铸造工艺，可以制造薄壁、形状复杂、花纹精细、工序繁杂的铜饰雕塑。制作中首先采用精铸工艺翻制石蜡模型，用石英砂等精铸材料制作雕塑型壳，然后烤制型壳脱蜡后再行烧制型壳，接着浇铸铜水，待铜水冷却后打碎型壳完成铜品制作。最后对铜品表面做着色、防腐以及其他处理，这样才能完成雕塑制作。精美考究的铜雕装饰在陶瓷的表面，立刻突显家具不凡的身份与格调。

IMAGE © DI GAO MEI JU
图片来自蒂高美居品牌

NATURAL COMFORT
自然舒适

NATURAL COMFORT

自然舒适

Return to rural home

归园田居

Suddenly we start to be tired of the city. We are not accustomed to the hustle and bustle, the calculation between people, especially the more disturbing air, water and food, making it more powerless. We don't want to change the world, and there is no such ability. But we can change ourselves and our way of life. If everyone in the city chooses a wish, perhaps half of the people want to have a house in the city where they live, and the other half might hope to leave the city one day. Where shall we go when leave the city? Of course we will go to more suitable rural areas and have a rural life.

"Two acres of dry farmland, one clay-tiled home, a cow, a dog, a cat, a pair of chickens; sunset time to rest, dig wells to drink water, farming for food, live a happy life." From the beginning of the development of industrial society, there has been a reflection of urban civilization to return to the pastoral, and now it is more popular. However, world is changing and the most traditional pattern of farmers' life is hard to find. There are little mood of real farmers among urban people: "simple heart with no pursuit of luxury and greed. Live a life without restlessness, physically tired but spiritually relaxed."

The prosperity of city is not the forefront of trend of life, but the return to the traditional pastoral life is the ideal pursuit. Many people have begun to make the original desire for pastoral life gradually become a reality. This is the purpose of the chapter: behind the idyllic dream, a new way of life rises. Now the process of urbanization is getting faster and faster, while the traditional poetic villages inadvertently decline. However, people's inner garden dream is more and more intense. Step on the soil, pull a few radishes, pick a few tomatoes, rub directly by hands, bite one, mouthful of natural fragrance... simple and almost monotonous behavior allows people never bored for a long time in the city cages. Field ridges, vegetable garden, the sun and the fragrance of soil deliver the ideal pastoral world.

So, the way of life in the poem "We open your window over garden and field. Talk about mulberry and hemp with our cups in our hands", "While picking asters under the Eastern fence, my gaze is upon the Southern mountain rests" has been re-picked up and re-interpreted. People also hope that in their small home in the reinforced concrete forest they can still have their own pieces of "paradise." In the context, the most important thing is the contentment and calm state of mind and no disturbance by the material life, the unity of internal and external as well as the spirit and material. People are afraid to lose the existing items, in fact, more like a reaction to the fear of loss of social status. Changing the fundamental habits of life in fact demands one's starting to consider way of thinking. One who has too much material thing is a beggar. Richness is a state of mind rather than appearance. You may appreciate the beauty of the world that can be seen everywhere when you know how to appreciate the beauty of a flower.

不知道从什么时候开始，突然厌倦城市。不习惯过分的喧嚣、人与人之间的算计。尤其是越来越让人不安的空气、水和食品，平添了更多的无能为力。不想去改变世界，也没有这个能力。但可以改变自己，以及自己的生活方式。如果让住在城市里的每个人选一个愿望，或许会有一半的人希望能在生活的城市里有一套商品房，而另一半可能会说，希望有一天能离开这个城市……离开城市去哪儿呢？当然是一个比城市更适合居住生活的田园乡村。

“两甲旱田，一楹瓦屋，一头牛，一条狗，一只猫，一对鸡；日入而息，凿井而饮，耕田而食，含哺而熙，鼓腹而游。”实际上从工业社会发展之初，便不断有反思都市文明、希望回归田园的文字出现，到现在更是蔚为大观。怕只怕是世事变化，最传统的农人生活的模式亦是难寻，而真正能有农人心境的都市人又有几个：“纯朴的心不奢求，不贪欲，过着无所不足、劳力而不劳心的安详生活。”

城市的繁华已然不是最前沿的生活潮流，回归传统的田园牧野生活才是最理想的追求，很多人也开始从当初对田园生活的渴望逐渐变成现实。这便是本章节的心声：田园梦背后，新生活方式崛起。现如今城市化进程越来越快，传统的诗意乡村在不经意间衰落，人们内心的田园梦却越来越强烈。脚踩泥土，拔几棵萝卜，摘几个西红柿，就地用手搓两下，啃一口，满嘴自然的清香……简单得近乎单调的行为中，却让久在都市牢笼里的人乐此不疲。田埂、菜园、阳光、泥土的芬芳，垒出的是理想的田园世界。

如此，“开轩面场圃，把酒话桑麻”“采菊南山下，悠然见南山”的生活方式得到了人们的再次捡拾和重新诠释，也希望能在处于钢筋水泥森林中的小家里拥有自己的一片“世外桃源”。而处于这一派情境之中，最重要的其实是能否拥有知足而平静的心境，能否达到不被物质所烦扰的目的，能否讲求内外在、精神与物质合一。人们害怕失去拥有的物品，其实更像是反映了对失去社会地位的恐惧，而要改变生活习惯的根本其实得先从思考方式开始。拥有太多的人才是乞丐，富有是一种心境而不是表象，当你懂得欣赏存在于一朵花中的美好，就可能学会欣赏世界中各处可见的美。

Bustling faded, fresh leisure and back to nature.

——繁华褪尽 清新休闲 返璞归真

IMAGE © GLOBAL VIEWS
图片来自 GLOBAL VIEWS 品牌

IMAGE © GLOBAL VIEWS
图片来自 GLOBAL VIEWS 品牌

Pastoral leisure

田园休闲牧歌

Modern living room in the pastoral style design advocates "return to nature". Only by combining nature, can we obtain a balance between physical and psychological in today's fast-paced social life. So the pastoral style showing the natural pastoral life just meets the people's concern for rapid expansion, urban environment deterioration, increasing estrangement between people etc.

After a busy day outside, we really want to head into our own home, and thoroughly enjoy the comfort of the recliner and the warmth of sunshine through the windows, and enjoy the fragrance of wood furniture distribution, and even a fireplace in the winter while watching a wonderful novel. British countryside respected by Lin Yutang is certainly a representative of pastoral style. British pastoral style was formed around the end of the 17th century, mainly due to people's boredom of the luxury style, they are longing for the fresh rustic style.

Feature

Floral pattern is the eternal main theme of English pastoral style. Furniture is mostly hand-fabric-based cloth, with beautiful lines and elegant colors. Ornament cloth is also adhering to the distinctive feature making it unforgetable.

Cloth

Beautiful and handmade cloth has pretty color and takes numerous flowers as the main pattern. There are also floral, stripes, Scotland lattice, and every kind of cloth is full of local flavor.

Material

English pastoral furniture often uses pine and piles. Its production and carving are all handmade and very particular about color. Furniture is mostly white, ivory and other white-based, with elegant shape, detailed lines and high-grade paint treatment, making every product like a mature and elegant middle-aged woman exuding calmness and elegance, but also like a 18-year-old girl who has pure and refined temperament, arousing continuous imagination.

American village belonged to the natural style can also meet expectation of "home", which gives you happiness, as if walking through the golden wheat field on your way back home, humming leisurely. American country style is pleasant, elegant and leisure in appearance, which advocates "return to nature", and strives to show leisure, comfortable, natural pastoral life.

Feature

American country style emphasizes rural comfort design criteria to pursue the original material sense and pay attention to comparison between roughness of the material itself and fineness of workmanship.

Color

Take elegant slate and antique white as main tone, and free graffiti floral pattern for the mainstream characteristics. Line is in free style but clean and distinct. Delicate and uniform color, gorgeous and low-key pattern are full of rich flavor of life.

Material

It is usually with simplified lines, rough size. The selection is also very broad: solid wood, printed cloth, hand-woven nylon material, linen fabrics and natural cutting stone... It often uses natural wood, stone, rattan, bamboo, red brick and other materials without polish, to show rustic texture; take practical-oriented principal to commonly use pine, oak and others to decorate to show a sense of old.

现代居室中的田园风格设计当然倡导“回归自然”，只有结合自然，才能在当今快节奏的社会生活中获取生理和心理的平衡。因此，田园风格力求表现自然的田园生活情趣。而这样的自然情趣正好处于现今人们对于人类城市扩张迅速、城市环境恶化、人们日渐互相产生隔阂而担心的时代。迎合了人们对于自然环境的关心，回归和渴望之情，所以也就造就了田园风格设计在当今时代的复兴和流行。

在外面忙碌了一天，真想一头扎进自己的家里，彻底地享受躺椅的舒适，享受透过窗户射进来的温暖阳光，享受原木家具散发的清香，甚至还该有个壁炉，冬天可以边烤火边看一本精彩的小说。被林语堂推崇的英伦乡村当然是田园风格里颇具代表性的一大类型。英式田园风格大约形成于 17 世纪末，主要是由于人们看腻了奢华风，转而向往清新的乡野风格。

特征

碎花图案是永恒的英式田园风格的主调，家具多以手工布面为主，线条优美，颜色雅致，饰品布艺也秉承了这个特点，特征鲜明让人过目不忘。

布艺

华美的布艺以及纯手工的制作，布面花色秀丽，多以纷繁的花卉图案为主。另外也有碎花、条纹、苏格兰格，每一种布艺都乡土味道十足。

材质

英式田园家具多使用松木、椿木，制作以及雕刻全是纯手工的，十分讲究配色。家具多以奶白、象牙白等白色为主，优雅的造型、细致的线条和高档油漆处理，都使得每一件产品像优雅成熟的中年女子含蓄温婉内敛而不张扬，散发着从容淡雅的生活气息，又宛若姑娘十八清纯脱俗的气质，无不让人心潮澎湃，浮想联翩。

而对“家”的期望，属于自然风格另一支的美式乡村亦可以满足，它给你的快乐，就像你回家时走在金色麦田路上、哼着悠然小调时所享受的快乐一样。美式乡村风格，清婉惬意，外观雅致休闲，倡导“回归自然”，力求表现悠闲、舒畅、自然的田园生活情趣。

特征

美式乡村风格以舒适为设计准则，更加追求材质的原始感觉，讲究材质本身的粗糙与做工的精细的对比。

色彩

以淡雅的板岩色和古董白居多，随意涂鸦的花卉图案为主流特色，线条随意但注重干净干练。细腻而统一的色调、华丽又低调的图案，充满了浓郁的生活气息。

材质

通常具备简化的线条、粗犷的体积，其选材也十分广泛：实木、印花布、手工纺织的尼料、麻织物以及自然裁切的石材……常运用不经雕琢的纯天然木、石、藤、竹、红砖等材质，展现质朴的纹理；以实用为主，常用松木、橡木等进行装饰，显出陈旧感。

8

FLOLENCO

佛洛伦克

IMAGE © FLOLENCO
图片来自佛洛伦克品牌

Robison Crusoe
Robison Crusoe
Robison Crusoe
LES MISERABLES
Count of Monte Cristo

IMAGE © FLOLENCO
图片来自佛洛伦克品牌

IMAGE © FLOLENCO
图片来自佛洛伦克品牌

8.1 LIVING ROOM
客厅赏析

Light blue fabric sofa matches the bright yellow leather sofa making the entire space light and stylish. Elm aging treatment bookcase without carving can still maintain the original texture and vein to bring more sense of layering for the whole space. Adding white elegant tea sets and marine style table lamp, we can experience complex romance and elegance, which embodies a beauty of style and expresses the comfortable beauty of life.

浅蓝的布艺沙发搭配明亮的黄色皮质沙发使整个空间轻盈而不失时尚。榆木做旧的书柜不用雕饰仍保有木材原始的纹理和质感，为整个空间带来更多层次感。加入白色优雅的茶具和海洋风格台灯，繁复中体会浪漫与优雅，体现的是一种格调美，亦表达生活惬意之美。

IMAGE © FLOLENCO
图片来自佛洛伦克品牌
The Three Musketeers
Horses

The series of living room has distinguished classic, calm temperament, and elegant colors, with both beautiful modeling and practical function. In the choice of material it is very critical. Solid wood of leather with the epidermis aging treatment exudes a low-key sense of luxury, reflecting the classical sense of elegance and utility of classical American style. In the use of color, with black, coffee and brown-based, the type of color in the home furnishing has good effects, dirtproof and easy to care. It is also elegant in the production of furniture. With high aesthetic and practical value, large furniture decorated with animal elements of the home accessories makes space more vivid.

这组客厅系列尊贵古典、沉稳大气、色彩典雅，兼具优美的造型与实用的功能。在材质的选择上十分考究，真皮搭配表皮做旧的实木散发出低调的奢华感，体现出古典美式的高贵感和实用。在用色上，以黑色、咖色和棕色为主，这类色在家居中搭配效果好，耐脏，且易于打理。在家具制作上工艺考究，极具审美价值和实用价值，大件家具点缀动物元素的家居饰品使空间更加生动活泼。

LES MISERABLES
LES MISERABLES
LES MISERABLES
LES MISERABLES
LES MISERABLES

SUCH
WELCOME
SUCH
FAREWELL

IMAGE © FLOLENCO
图片来自佛洛伦克品牌

This set of living room series is fashion and avant-garde, and its material and color are novel and bold. Black and white colorlessness embellishes bright orange to make the space more fresh and playful. Products abandon too many complicated decorations, and adopt natural concise lines to create a stylish and modern atmosphere. High contrast between high-reflective stainless steel and black & white stripes wallpaper complements each other, and vivid modeling ornaments make space more interesting.

这组客厅系列时尚前卫，材质和配色新颖大胆，黑白无彩色点缀鲜艳的橙色让空间更加鲜活俏皮。产品摒弃了过多的繁杂装饰，自然简洁的线条营造时尚现代的气氛，不锈钢的高反光度与黑白条纹壁纸的高对比度相得益彰，造型生动的饰品摆件让空间更加有趣。

The series of dining table is exquisite and fashionable with strong visual impact. It continues composition elements and spatial characteristics of the living room, with bright and bold colors as well as avant-garde and playful features. Coral shells on the table are filled with the breath of life. Wine-glasses chandeliers are full of creation and reflect the dazzling light in the light irradiation.

这组餐桌系列精致时尚，视觉冲击力强，延续客厅的组成元素和空间特点，用色鲜艳大胆，前卫俏皮。餐桌上面的贝壳珊瑚富于生活的气息，酒杯吊灯创意十足，在灯光的照射下反射出耀眼的光芒。

8.2 DINING ROOM
餐厅赏析

IMAGE © FLOLENCO
图片来自佛洛伦克品牌

SUCCESS
IS THE
BEST
REVENGE
K

8.3 PRODUCT DISPLAY
单品赏析

IMAGE © FLOLENCO
图片来自佛洛伦克品牌

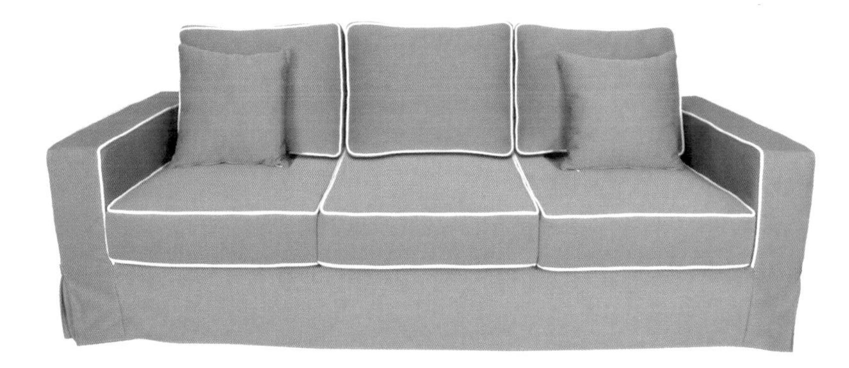

01 Shiraz three-seat sofa
Size l 215cm×90cm×90cm
Material l Linen cloth

施拉三人位
尺寸 | 215cm×90cm×90cm
材料 | 棉麻布料

02 Taichi bookcase
Size l 112cm×42cm×220cm
Material l Oak, Stainless steel

泰奇书柜
尺寸 | 112cm×42cm×220cm
材料 | 橡木，不锈钢

03 Tree branches Candlestic
Size l 32cm×15cm×53cn
Material l Iron, Resin

树枝烛台 B
尺寸 | 32cm×15cm×53cm
材料 | 铁艺，树脂

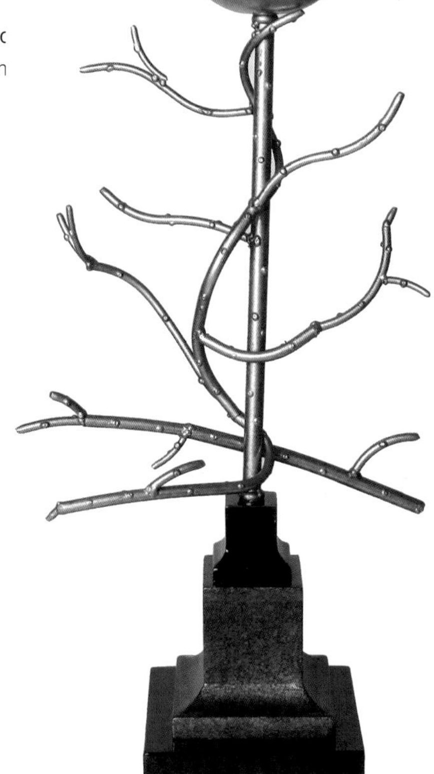

04 Pinos Decorative Tank A / B
Size l 20cm×20cm×42cm / 28cm
Material l High Temperature Ceramics

皮诺斯装饰罐 A/B
尺寸 | 20cm×20cm×42cm / 28cm
材料 | 高温陶瓷

05 Electronic Candle — Orange A / B
Size I 7.5cm × 12.5cm / 9cm
Material I Built-in 5 batterie

电子蜡烛——橙色 A/B
尺寸 | 7.5×12.5cm/9cm
材料 | 内装 5 号电池

06 Black pallet
Size I 38cm × 48cm
Material I Wood, glass

黑色托盘
尺寸 | 38cm×48cm
材料 | 木，玻璃

07 Onyx Table Lamps — Red
Size I Lamp body 21cm × 10cm × 48cm
cover 35cm × 35cm × 26cm
Material I Agate tablets (Natural materials, agate shape of each product is different, and there are color differences)

玛瑙台灯——红色
尺寸 | 灯体 21cm×10cm×48cm
罩 35cm×35cm×26cm
材料 | 玛瑙片（天然材料，每个产品玛瑙片形状不同，且有色差）

08 Onyx Table Lamps — Blue
Size I Lamp body 21cm × 10cm × 48cm
cover 35cm × 35cm × 26cm
Material I Agate tablets (Natural materials, agate shape of each product is different, and there are color differences)

玛瑙台灯——蓝色
尺寸 | 灯体 21cm×10cm×48cm
罩 35cm×35cm×26cm
材料 | 玛瑙片（天然材料，每个产品玛瑙片形状不同，且有色差）

09 High-back chair — Red
Size I 42cm × 50cm × 107cm
Material I Oak, Thick fabric, Imitation leather

高背椅——红色
尺寸 | 42cm×50cm×107cm
材料 | 柞木，超厚布料，仿皮

IMAGE © FLOLENCO
图片来自佛洛伦克品牌

10 Peiqi convex mirror

Size I 52cm×52cm×8cm
Material I MDF, Convex mirror

佩琪凸镜

尺寸 | 52cm×52cm×8cm
材料 | MDF，凸镜

11 Spar decorative painting D

Size I 40cm×40cm×5cm
Material I Solid wood frame, Natural ore

晶石装饰画 D

尺寸 | 40cm×40cm×5cm
材料 | 实木框，天然矿石

12 Ancient Greek lamp

Size I 85cm×45cm
Material I Aluminum Plating Nickel

古希台灯

尺寸 | 85cm×45cm
材料 | 铝镀镍

13 NISI Triple — Orange Materials

Size I 215cm×90cm×80cm
Material I Linen cloth

尼斯三人位——橙色

尺寸 | 215cm×90cm×80cm
材料 | 棉麻布料

14 Ike cans — red AB

Size I 15.5cm×15.5cm×48cm/
14.5cm×14.5cm×42cm
Material I High temperature ceramics

艾克罐——红色 AB

尺寸 | 15.5cm×15.5cm×48cm/
14.5cm×14.5cm×42cm
材料 | 高温陶瓷

IMAGE © FLOLENCO
图片来自佛洛伦克品牌

15 Blue ripple lamp

Material l High temperature ceramics, transparent base

蓝色波纹台灯

材料 | 高温陶瓷，透明底座

16 Shells and Western food sets

Material l Copper alloys

贝壳西餐套

材料 | 铜合金

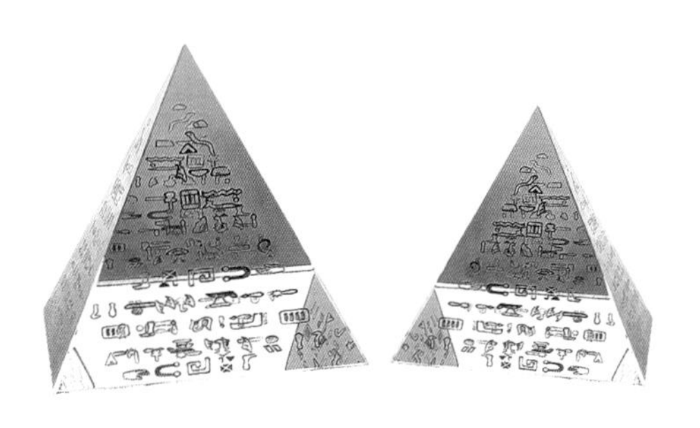

17 Crystal Pyramid AB

Size l 10cm×10cm×11cm/8cm×9cm
Material l K9 crystal material

水晶金字塔 AB

尺寸 | 10cm×10cm×11cm/8cm×9cm
材料 | K9 水晶料

18 XINUO tea sets

Material l Copper

希诺茶具套

材料 | 铜

19 Danny picture frame AB

Size l 15cm×4.38cm×20cm/12.5cm×4.38cm×15cm
Material l Aluminum plating copper

丹尼相框 AB

尺寸 | 15cm×4.38cm×20cm/12.5cm×4.38cm×15cm
材料 | 铝镀铜

20 Paul barrels A

Size l 25cm×15cm×20cm
Material l Brass with nickel plating

保罗酒桶 A

尺寸 | 25cm×15cm×20cm
材料 | 黄铜镀镍

21 Darth vases
Size l 25cm×39.5cm
Material l Copper Nickel Plated, Genuine leather

达斯花瓶
尺寸 | 25cm×39.5cm
材料 | 铜镀镍，真皮

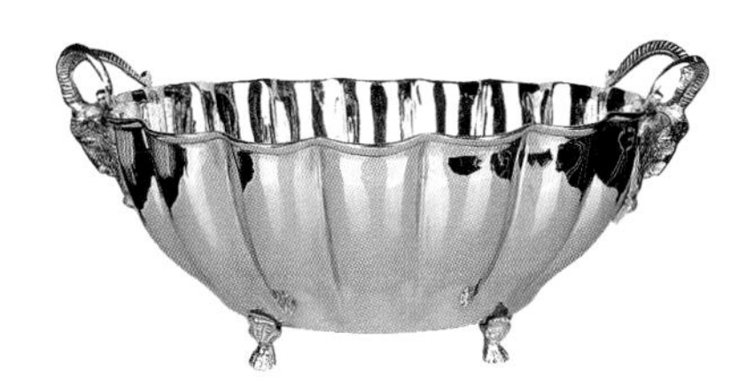

22 Orchid Decoration A
Size l 30cm×30cm×30cm
Material l Brass plate plated in light gold

兰度装饰 A
尺寸 | 30cm×30cm×30cm
材料 | 黄铜片镀浅金色

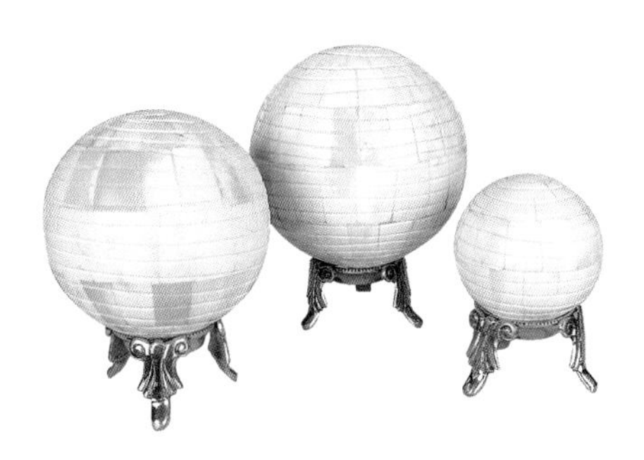

23 Jamie decorative ball AB
Size l 13cm×17cm /10×14cm
Material l Cow bone, Brass base

杰米装饰球 AB
尺寸 | 13cm×17cm/10cm×14cm
材料 | 牛骨，黄铜底座

24 Shells petals napkin clasp
Material l Shell handmade string

贝壳花瓣餐巾扣
材料 | 贝壳手工串

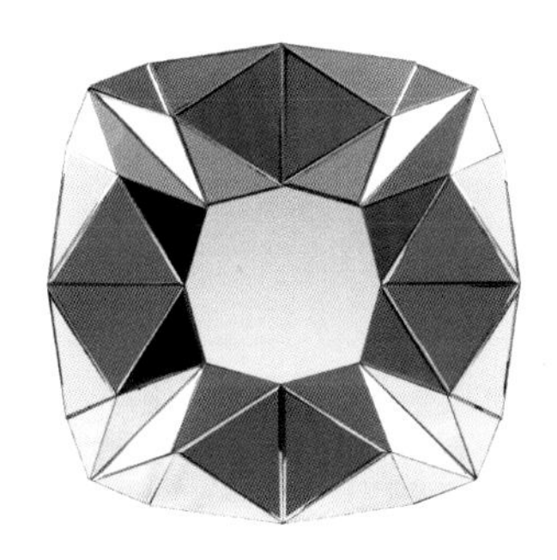

25 Mogh mirror
Size l 76.2cm×76.2cm×5.1cm
Material l MDF, Mirror

摩格镜
尺寸 | 76.2cm×76.2cm×5.1cm
材料 | MDF，镜子

26 Clay tea table — silver
Size l 63.75cm×36.88cm×52.5cm
Material l Aluminum, Copper

克莱边几——银色
尺寸 | 63.75cm×36.88cm×52.5cm
材料 | 铝，铜

IMAGE © FLOLENCO
图片来自佛洛伦克品牌

27 LIKE entrance frame

Simple and neat materials like brass and glass, with square and concise profile shape and lines, modernization transformation of the classic shape, material, which is not exaggerated, but to create a subtle sense of fantasy, let everything orderly.

里可玄关架

黄铜、玻璃这样质朴利落的材料，配合方正简洁的廓形和线条，成功将经典造型、材质进行现代化改造的作品，很实在一点都不浮夸，却营造出了一种微妙的奇幻感，让一切显得秩序井然。

28 Leather baskets

As if transfigure the old cortex into practical container, emitting a distant sense of the era, enriching monotonous indoor, increasing the sense of accumulation of space, revealing the durability and perseverance of time.

皮篓

仿佛将老式的旧皮质变身实用的置物容器，散发出遥远的年代感，丰富单调的室内，增加空间的积淀感，透露着经得住时间打磨的耐用与恒心。

29 England carpet

Scottish Check is the most representative check pattern of the British style. The classic check has been given modern interpretation by designer, with more complex lines and check as well as the hand-woven texture.

英格地毯

苏格兰格纹是最能代表英伦风格的格纹图案，这些经典格纹已被设计师赋予了现代化的演绎，拥有更加复杂的格纹线条和纯手工的编织肌理。

31 Scallop Decorations

Artificial alloy and natural shell in integration turn into unique creative handmade ornaments, just like back to the beach to listen to sea breeze blowing and the waves roaring...

扇壳摆件

人工的合金与天然的贝壳相融合，幻化为独特的创意手工摆件，感觉似又回到了海风吹拂的沙滩上，听浪潮波涛……

30 Sini Box Coffee Tables

With aluminum skin, MDF fiberboard and simulation skin as the material, it becomes old box coffee tables for nostalgia people. Nostalgic elements create a thick old color, with a tough temperament to interpret the retro style of industry.

斯尼箱子茶几

以铝皮、MDF 纤维板与仿真皮为材料，为恋旧一族组合而成大箱子茶几，怀旧元素打造出厚重的旧物本色，用一种硬朗的气质去诠释复古工业之风。

33 Black and white striped chairs
Size I 49cm×57cm×94cm
Material I Beech

黑白条纹椅

尺寸 | 49cm×57cm×94cm
材料 | 榉木

32 Aspen table lamp — black
Size I 84cm×50cm
Material I Copper Nickel Plated

艾斯台灯——黑

尺寸 | 84cm×50cm
材料 | 铜镀镍

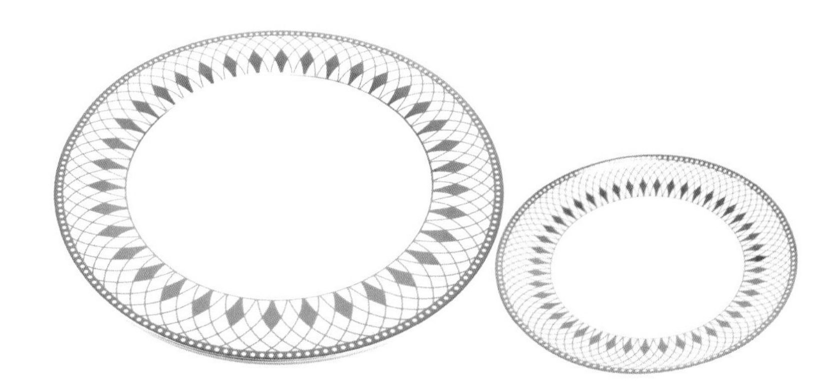

34 Checkered plate A/B
Size I 28cm×28cm×2.5cm/
19cm×19cm×2cm
Material I Bone china

格纹餐盘 A/B

尺寸 | 28cm×28cm×2.5cm/
19cm×19cm×2cm
材料 | 骨瓷

IMAGE © FLOLENCO
图片来自佛洛伦克品牌

35 Mis three-seat sofa

Size I 229cm×99cm×80cm
Material I Oil wax or Imitation leather

米斯三人位

尺寸 | 229×99×80CM
材料 | 油蜡皮或仿皮

36 Fur photo frame

Import
Size I 19cm×25cm
Material I Hand rub color and aging treatment cattle leather

皮相框

进口
尺寸 | 19cm×25cm
材料 | 手擦色做旧牛皮

37 Sino decoration A/B/C

Import
Size I 15cm×15cm×50cm/13cm×13cm×42cm/11cm×11cm×35cm
Material I Aluminium, Cattle bone

西诺装饰 A/B/C

进口
尺寸 | 15cm×15cm×50cm/13cm×13cm×42cm/11cm×11cm×35cm
材料 | 铝，牛骨

IMAGE © FLOLENCO
图片来自佛洛伦克品牌

38 Cattle leather round tray
Import
Size l 40cm×40cm×5cm
Material l Hand rub color and aging treatment cattle leather

牛皮圆托盘
进口
尺寸 | 40cm×40cm×5cm
材料 | 手擦色做旧牛皮

40 The zebra pillow
Import
Size l 45cm×45cm
Material l Wool, Cotton

斑马纹抱枕
进口
尺寸 | 45cm×45cm
材料 | 羊毛，棉

39 Leopard Dog AB
Size l 24.5cm×10cm×22cm/13.5cm×9cm×20cm
Material l Resin Copper Plating

豹犬 AB
尺寸 | 24.5cm×10cm×22cm/13.5cm×9cm×20cm
材料 | 树脂镀铜

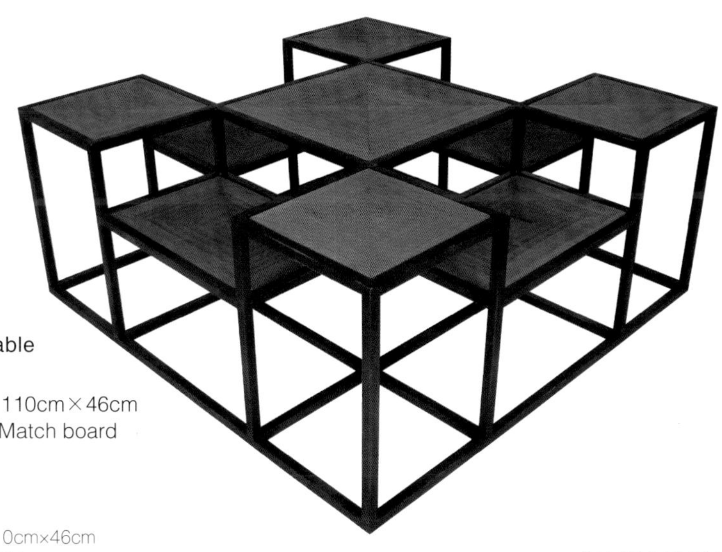

41 North coffee table
Import
Size l 110cm×110cm×46cm
Material l Iron, Match board

诺斯茶几
进口
尺寸 | 110cm×110cm×46cm
材料 | 铁艺，拼花板

IMAGE © FLOLENCO
图片来自佛洛伦克品牌

42 Lenny sofa

Size I 64cm×70cm×92cm
Material I Imitation leather or oil wax skin

莱尼沙发

尺寸 | 64cm×70cm×92cm
材料 | 仿皮或油蜡皮

43 Cox three-seat sofa

Size I 210cm×88cm×110cm
Material I Linen cloth

库斯三人位

尺寸 | 210cm×88cm×110cm
材料 | 棉麻布料

44 SINY Box coffee table

Size I 120cm×70cm×50cm
Material I Aluminum skin, MDF, Simulation leather

斯尼箱子茶几

尺寸 | 120cm×70cm×50cm
材料 | 铝皮，MDF，仿真皮

45 SINY six drawers cabinet

Size I 60cm×50cm×120cm
Material I Aluminum skin, MDF, Simulation leather

斯尼六抽柜

尺寸 | 60cm×50cm×120cm
材料 | 铝皮，MDF，仿真皮

46 SIQI three-seat sofa
Size | 228cm × 92cm × 95cm
Material | Linen cloth

思琪三人位

尺寸 | 228cm×92cm×95cm
材料 | 亚麻布料

47 Crystal ball decoration A/C
Size | 26cm × 35cm/18cm × 27cm
Material | Glass

水晶球装饰 A/C

尺寸 | 26cm×35cm/18cm×27cm
材料 | 玻璃

48 LIKE long coffee table
Size | 111cm × 61cm × 55cm
Material | Brass

里可长茶几

尺寸 | 111cm×61cm×55cm
材料 | 黄铜

IMAGE © FLOLENCO
图片来自佛洛伦克品牌

BRAND SPONSORS

品牌赞助商

GLOBAL VIEWS

Global Views was founded in 1997 in Dallas, US. The main business is global high-end products and high-end space design and implementation. Brand founder David Gebhart spent 15 years on in-depth understanding of product design and customer preference, paving the way for Global Views' many years of retail procurement experience, strong product development capabilities, and professional marketing knowledge from the very beginning of the establishment.

GLOBAL VIEWS 于 1997 年在美国达拉斯创立，主力经营全球家居高端产品以及高档空间设计和实现。品牌创始人 David Gebhart 在此前用了 15 年的时间深入了解产品设计和客户喜好，使得 GLOBAL VIEWS 从创立伊始便已具备多年零售采购的丰富经验，强大的产品开发能力，以及专业的市场营销知识。GLOBAL VIEWS 作为装饰、设计和色彩潮流的领头羊，以对家居时尚的创新性思维而闻名，致力于将最好的家具和家居饰品带给市场。

欧米亚

"Stainless steel household goods experts" is the core concept of Omenia brand. Omenia advocates minimalism and modern art, and the main material of products is stainless steel, using extensive of environmentally friendly metal materials. The products include creative furniture, art decoration, tea sets, kitchen supplies, and hotel supplies and so on.

"不锈钢家居用品专家"是欧米亚的品牌核心理念。欧米亚崇尚极简主义和现代艺术，产品以不锈钢为主要材料，大量运用各种环保金属材质，生产的产品包括创意家具、艺术摆件、茶具套件、厨房用品、酒店用品等。

帷澜

Guangzhou Weland Handicraft Co., Ltd. was founded in 2007, with brands AMES, Luxor and EXP +. Based on the Pearl River Delta, Weland is characterized by the integration of handicraft sales and space hard and soft decoration program. Different from the strong visual design of modern home furniture, Weland advocats lifestyle focusing on user experience and creating a living atmosphere. Weland is committed to making living space conducive to people's daily communication and emotional sublimation. It is to restore life and beautify life, rather than a fine but not practical, nice but not warm house.

广州市帷澜工艺品有限公司(Weland)成立于2007年，现旗下品牌有爱美仕（AMES）、乐斯（Luxor）与 EXP+。立足珠三角，工艺品的研发销售与空间硬软装搭配方案一体化是其特点。区别于现代家居强烈的视觉设计，Weland 主张倡导的生活方式，更倾向于注重用户体验和营造居住氛围。Weland 致力于把居住空间打造成有利于人们日常交流、升华情感的地方，它还原生活，美化生活，而不是一所精致但不实际，好看但无温度的房子。

自在工坊

Ease workshop was founded in 2011. Following the idea of "the example of nature as a teacher, free symbiotic", and through the combination of traditional and modern technology, as well as the use of diversified materials, it adapts to the contemporary life of people, objects, space relations, so to achieve symbiosis between people and society, people and nature. In retrospect to Oriental Zen aesthetic.and integration of modern Western design concepts, it is committed to creating the modern Oriental home brand blending natural, humanities, design in one. It has won several awards at home and abroad.

自在工坊成立于 2011 年，遵循"师法自然，自在共生"的理念。通过传统与现代工艺相结合，多元化材料运用，建立适应当代生活的"人、物、空间"关系，实现人与社会，人与自然共生。追溯东方禅意美学，融合西方现代设计理念，致力打造融自然、人文、设计于一体的现代东方家居品牌，曾多次获得海内外奖项。

春在东方

CZDF incorporates research and development, production, sales and soft decoration design in one, to provide customers with "Neo-Chinese style" cultural home accessories one-stop overall soft decoration services. With the keen sense of Chinese culture and home furnishing trends, CZDF forms an "international, fashional, Oriental" research and development team, and takes the lead in proposing the "Neo-Chinese style" concept of the overall home furnishing. Products cover ceramics, lighting, decorative painting, wood, screens, furniture, sculpture, wallpaper and other home furnishing accessories.

Its household products have five major brand series and two design agencies. The brand "Hua Tian Imagination" focuses on the design of Neo-Chinese style commercial space, "Dong Yun Soft Decoration Organization" focuses on the design and implementation of Neo-Chinese style soft decoration.

CZDF inherits the Eastern classics, leading cultural rejuvenation and integrating the essence of different cultures and times while interpreting the Oriental temperament.

春在东方集研发、生产、销售及软装设计于一体，为客户提供"新中国风"文化家居饰品一站式整体软装服务。凭借对中国文化及家居潮流的敏锐嗅觉，组建了"国际的、时尚的、东方的"研发设计团队，率先提出"新中国风"整体家居的概念。产品涵盖陶瓷、灯饰、装饰画、木器、屏风、家具、雕塑、墙纸等各类家居饰品。

旗下家居产品有五大品牌系列、两个设计机构，其中"华天创想"专注新中国风商业空间设计、"东韵软装机构"专注新中国风软装陈设设计及实施。

春在东方传承东方经典，引领文化新生，融汇不同文化与时代的精粹，诠释东方气质与内涵。

铂晶艺术

BOKING Art Studio is a group of persistent artists. Studio's research direction is the unique platinum crystal transparent material as the core to original artistic creation. After years of independent research and development, BOKING wins a major technological breakthrough in the field of materials with its high transparent material molding process. The spirit of perseverance promotes them to continue to break through the limit and create a miracle repeatedly.

Daijie Wu, founder of the studio, considers that contemporary art of China needs innovation, but more in need of historical heritage. Ideal works of art should have sufficient visual beauty to show the existence of human beings as well as a certain philosophical thinking and inspirations.

These pure and bright works are different from today's parrot art form. Whether the theme or style of the work is showing a very unique form of pure personality.

It is commendable that the studio has been in the pursuit of self-character and has been devoted in their work around the same material and technology. This kind of material is transparent as crystals, which can be made in various shapes. BOKING is very fascinated by the material.

铂晶艺术工作室是由一群执着的艺术家团队组合而成，工作室研究的方向是以特有的铂晶透明材质为核心进行原创艺术创作。工作室历经多年自主研发的高透明材料成型工艺是材料领域的重大技术突破。执着的精神使他们不断突破极限、屡创奇迹。

工作室的创始人认为中国的当代艺术需要创新，但却更需要历史的传承。理想的艺术作品应该是具有足够的视觉美感、能够表现人类的生存状态以及具有一定的哲学思想，并需要一定的启示性。

这些纯净璀璨的作品的确与当今那些人云亦云的艺术面貌拉开了距离，无论是作品的主题抑或是艺术的风格，都呈现出独特的非常纯粹的个性化形态。

难能可贵的是，工作室一直在追求自我的本色，并且一直围绕同一种材质和工艺来实现自己的作品，这种材质如同水晶一般透明，却可以成型很丰富，工作室对这种材质非常着迷。

蒂高美居

As a model of luxury home furnishing — DI GAO MEI JU not only inherits the classic charm of Chinese and foreign classical home accessories in design, but also provides products perfectly in match with fashionable jewelry trend, making classical and modern, art and practice perfect unity. The main products of DI GAO MEI JU include high-grade European and American classical ceramics, luxury copper with porcelain home accessories and exquisite hand-painted decorative small furniture. Products bring European and American luxury jewelry a new trend with rare top-quality material and superb craftsmanship. And extensive use of precious metals such as gold, silver, copper and superior porcelain creates a magnificent luxury expression for the jewelry industry. And by increasing the sense of history, in addition to appreciation it also has maintenance, collection and other value.

DI GAO MEI JU shows retro and classic vividly. With elegant shape, exquisite sculpture, European-style modeling, every product is like a piece of artwork, and a feast to the eyes for all people.

作为奢华美居饰品的典范——"蒂高美居"在设计中不仅传承了中外古典精品家居饰品的经典韵味，同时产品亦十分贴合风尚饰品潮流，将古典与现代、艺术与实用完美统一。"蒂高美居"主要产品包括高档的欧美古典陶瓷、奢华的铜配瓷家居饰品和精美的手绘装饰小家具，产品以稀有的顶级材质、精湛的传世手工艺引领了欧美奢华美居饰品新潮流，并大量运用了金银铜等贵金属及上好的瓷土，为家居饰品行业打造出金碧辉煌的奢华表情，同时增加历史的厚重，使家居饰品除鉴赏外还具有保值、收藏等价值。

"蒂高美居"把复古与经典展现得淋漓尽致，每一件产品都像一个艺术品，优雅的造型、精美的雕刻、欧式的造型，无一不让人赏心悦目。

佛洛伦克

FLOLENCO Group is a high-end overall home products supplier focusing on the design, production, sale and display of home furnishing products. The brand "FLOLENCO" has been advocating low-key, elegant, tasteful and natural overseas home culture for many years. The company has built up a highly personalized and international art life space for clients, and won many domestic awards and high popularity in home furnishing industry. Professional international design team shuttles every year in Paris, Milan, High Point (North Carolina) and other world's major home furniture exhibitions to collect the latest and most cutting-edge trend of home furnishing. In October 2014, the official opening of the Indian branch, helps to process a new level for the global integration of FLOLENCO boutique home furnishing. In the future FLOLENCO will continue to follow the core values of "sincerity", "quality" and "innovation" and make continuous efforts to create an internationalized home furnishing culture atmosphere and further spread brand culture and fine art life space.

佛洛伦克集团是专注于家居产品的设计、生产、销售和陈设为一体的中高端整体家居产品供应商。旗下品牌"FLOLENCO 佛洛伦克"多年来坚持倡导低调、优雅、品位、自然的海外家居文化，用心为客户打造极具个性的、国际化的艺术生活空间，多次获得国内奖项，在行业享有较高的知名度。专业的国际化设计团队，每年都穿梭于法国巴黎、意大利米兰、美国高点等世界各大家居展会收集最新、最前沿的家居潮流趋势。2014 年 10 月印度分公司正式挂牌运营，为佛洛伦克全球范围整合精品家居的进程迈上了一个新的台阶。身处行业的朝阳时代，佛洛伦克人将继续遵循"真诚""品质""创新"的核心价值观，为营造国际化的家居文化氛围和传播品牌文化、美好艺术生活空间不断努力。

CONTRIBUTORS

设计师名录

Yan Xichao　　闫希超

Mr. Yan advocates the discovery of beautiful things from life. A plant, a building, a piece of jewelry, clothing, and so on, are the source of inspiration, extending his intricate design. Since the creation of Omenia in 2005, he has always been committed to the development of soft ornaments in China.

Mr. Yan brings for soft-decoration design, hotel engineering and other fields a novel fashion light luxury experience since entering the metal home furnishing industry. Diversified style design, emphasizing the balance between personality and practical, applies the forefront of fashion metal jewelry to various fields. His design is the integration of modern urban life attitude and manner, the perfect combination of fashion and furnishings.

闫希超先生主张从生活中发现美丽事物。一株植物、一栋建筑、一件首饰、一件服装等，都是设计灵感来源，延伸出其巧夺天工的设计。自 2005 年创立欧米亚十余年来，始终致力于中国软装饰品发展。

闫先生自踏入金属家居业伊始，便为各类软装设计师、酒店工程等领域带来一种新颖时尚的轻奢体验。多元化的风格设计，强调个性与实用的平衡，将时尚前沿的金属饰品应用到各个领域。融合现代都市人的生活态度和方式，完美地将时尚与陈设结合为一体。

Zhang Jun　　张　军

Zhang Jun, born in Anhui Province in 1976, is major in Chinese classical garden, interior decoration and furniture design. He is an executive vice-chairman of Shenzhen Industry Design Association and executive chairman of Shenzhen Home Furnishing Design Committee. He has led industry innovation with design to promote the development of the industry. In 2014, he was invited to be a visiting professor of Central South University of Forestry and Technology. He is committed to bringing new ideas and frontline experience to students and promoting the development of China's furniture industry. In 2003 and 2011, he created the extension of Top Design and the original modern oriental home furnishing brand — Ease Workshop.

General Manager of Shenzhen Top Design Furniture Co., Ltd;
Vice-chairman of Shenzhen Industry Design Association;
Chairman of Shenzhen Furnishing Design Association;
Deputy Director of the Designers Professional Committee of Shenzhen Furniture Industry Association;
China Furniture top ten professional managers.

张军，1976 年生于安徽，学研中国古典园林、室内与家具设计专业，曾先后出任深圳工业协会常务副会长，深圳市家居设计专委会执行主席，以设计力量带领行业革新，推动行业发展。2014 年受聘中南林业科技大学客座教授，致力于将新的理念和一线经验传授于家具行业莘莘学子，校企联合，推动中国家具行业发展。于 2003 年和 2011 年先后创办拓璞设计及原创现代东方家居品牌自在工坊。

深圳市拓璞设计有限公司总经理；
深圳工业设计行业协会副会长；
深圳市家居设计专委会主席；
深圳市家具行业协会设计师专业委员会副主任；
中国家具十佳职业经理人。

ARTPOWER

Acknowledgements

We would like to thank all the designers and companies who made significant contributions to the compilation of this book. Without them, this project would not have been possible. We would also like to thank many others whose names did not appear on the credits, but made specific input and support for the project from beginning to end.

Future Editions

If you would like to contribute to the next edition of Artpower, please email us your details to: artpower@artpower.com.cn